T0235666

SEX AND GENDER IN PALEOPATHOLOGICAL PERSPECTIVE

A growing body of literature indicates that diseases can affect women and men differently. As sex differences extend far beyond biology, it is crucial to adopt a biocultural approach towards understanding human disease patterns and processes. This book synthesizes modern medical research with paleopathological investigations. Conditions such as osteoporosis and osteopenia, iron deficiency anemia, infection, immune reactivity, and trauma are explored. Recognizing the relationship between these conditions and aspects of sex and gender in past populations assists in the formulation of models from which modern disease processes can be better understood. Sex and Gender in Poleopathological Perspective will provide provocative ideas for all those in physical anthropology, archaeology, evolutionary biology, history of medicine, and women's studies who are interested in how sex and gender impacts on disease.

ANNE L. GRAUER is an Associate Professor of Anthropology in the Department of Sociology and Anthropology at Loyola University of Chicago. In 1993, she was awarded the Presidential Faculty Fellowship by the National Science Foundation. Her current research interests focus on women's health and social conditions in Medieval Britain and in the nineteenth-century United States. She has edited a book entitled *Bodies of Evidence: Reconstructing history through skeletal analyses* (1995).

PATRICIA STUART-MACADAM is an Associate Professor in the Anthropology Department of the University of Toronto. Her current research interests involve breast-feeding, trauma and musculoskeletal markers, sex and gender differences in disease, and iron deficiency anemia. She has co-edited two books, *Diet, Demography and Disease: Changing perspectives on anemia* (1992) with S. Kent, and *Breast-feeding: Biocultural perspectives* (1995), with K. Dettwyler.

SEX AND GENDER IN PALEOPATHOLOGICAL PERSPECTIVE

Edited by
ANNE L. GRAUER
PATRICIA STUART-MACADAM

CAMBRIDGE
UNIVERSITY PRESS

CAMBRIDGE UNIVERSITY PRESS
Cambridge, New York, Melbourne, Madrid, Cape Town, Singapore, São Paulo

Cambridge University Press
The Edinburgh Building, Cambridge CB2 2RU, UK

Published in the United States of America by Cambridge University Press, New York

www.cambridge.org
Information on this title: www.cambridge.org/9780521620901

© Cambridge University Press 1998

This publication is in copyright. Subject to statutory exception
and to the provisions of relevant collective licensing agreements,
no reproduction of any part may take place without
the written permission of Cambridge University Press.

First published 1998
This digitally printed first paperback version 2005

A catalogue record for this publication is available from the British Library

Library of Congress Cataloguing in Publication data

Exploring the differences : sex and gender in paleopathological
    perspective / edited by Anne L. Grauer, Patricia Stuart-Macadam.
        p.   cm.
    Includes index.
    ISBN 0 521 62090 2
    1. Sex factors in disease.   2. Sex differences.
    3. Paleopathology.   I. Grauer, Anne L.   II. Stuart-Macadam,
    Patricia, 1951–
    RB212.E96   1998
    616.07 – dc21   98-26485 CIP

ISBN-13 978-0-521-62090-1 hardback
ISBN-10 0-521-62090-2 hardback

ISBN-13 978-0-521-02121-0 paperback
ISBN-10 0-521-02121-9 paperback

To my family for their encouragement and support
(ALG)
To my children, Leila, Annie, Jamie, and James; living proof of the differences between the sexes and a very special thanks to James Macadam for his unfailing love and support
(PSM)

Contents

Contributors

George J. Armelagos
Department of Anthropology, Emory University, Atlanta, GA 30322, USA

Philip Boocock
Calvin Wells Laboratory, Department of Archaeological Sciences, University of Bradford, Bradford, West Yorkshire BD7 1DP, UK

Della Collins Cook
Department of Anthropology, University of Indiana, Bloomington, IN 47505, USA

Anne L. Grauer
Department of Sociology and Anthropology, Loyola University of Chicago, Chicago, IL 60626, USA

Diane V. Houdek
Chapin Hall, University of Chicago, Chicago, IL 60637, USA

Kevin D. Hunt
Department of Anthropology, University of Indiana, Bloomington, IN 47505, USA

Robert Jurmain
Department of Anthropology, San Jose State University, San Jose, CA 95192-0113, USA

Lynn Kilgore
Department of Anthropology, Colorado State University, Fort Collins, CO 80523, USA

Clark Spencer Larsen
Department of Anthropology and Research Laboratories of Archaeology, University of North Carolina, Chapel Hill, NC 27599-3120, USA

Thomas L. Leatherman
Department of Anthropology, University of South Carolina, Columbia, SC 29208, USA

Mary E. Lewis
Calvin Wells Laboratory, Department of Archaeological Sciences, University of Bradford, Bradford, West Yorkshire BD7 1DP, UK

Elizabeth M. McNamara
Department of Anthropology, University of Illinois, Urbana, IL 61801, USA

Donald J. Ortner
Department of Anthropology, National Museum of Natural History, Smithsonian Institution, Washington, DC 20560, USA

Charlotte A. Roberts
Calvin Wells Laboratory, Department of Archaeological Sciences, University of Bradford, Bradford, West Yorkshire BD7 1DP, UK

Rebecca Storey
Department of Anthropology, University of Houston, Houston, TX 77204–5882, USA

Patricia Stuart-Macadam
Department of Anthropology, University of Toronto, Toronto, ON M5S 1A1, Canada

David S. Weaver
Department of Anthropology, Wake Forest University, Winston-Salem, NC 27109, USA

Acknowledgments

I wish to extend my sincere thanks and appreciation to the contributors for their cooperation and patience. A special thanks goes to my current and former students Michele Buzon, Sherry Hanson, Patrick Waldron, Philip Hoog, and Theresa Jolly for their assistance in creating the index and for coping with innumerable manuscript details. Thanks also to Dr. R. A. Stuart for his comments, to Loyola University for providing resources, and to Dr. T. A. Sanderson, Commissioning Editor, Biological Sciences, at Cambridge University Press for fielding hundreds and hundreds of questions. This book is based upon work supported by the National Science Foundation under Grant No. SBR-9350256. Any opinions, findings, conclusions or recommendations expressed in this material are those of the authors and do not necessarily reflect the views of the National Science Foundation.
Anne L. Grauer

I wish to thank all the contributors for their participation in the 1995 AAPA symposium that I organized with Anne Grauer, which provided the stimulus for this book. Thank you to Dr. R. A. Stuart for his welcome editorial assistance. Thanks also to Dr Tracey Sanderson for believing in this book and being so enjoyable to work with. Lastly, I would like to give a special thanks to Anne Grauer, for her grace under fire and for holding it all together while I escaped 'down under,' and to James for always being there for me.
Patricia Stuart-Macadam

1

Introduction: sex, gender and health status in prehistoric and contemporary populations

GEORGE J. ARMELAGOS

Sex and gender remain concepts wrought with confusion. Even after three decades in which anthropologists have clarified the distinctions between sex and gender, confusion remains. There is a consensus in anthropology that sex is defined by the biological differences between males and females determined at the moment of conception and enhanced in subsequent physiological development. Sexual differences include features of the chromosome, genitalia, and other anatomical structures related to secondary sexual development. There is also agreement that gender is the cultural construct in which individuals are socially classified into categories such as male and female, or masculine and feminine in our culture. Other cultural systems recognize more than two gender classes. As anthropologists clarify the distinctions between the concepts, other disciplines are increasingly substituting the term 'gender' for the term 'sex'. Pearson (1996), a biologist, notes that the substitution of these terms reflects an attempt at political correctness that 'clouds understanding'. In response to Pearson's communication, Carlin (1996:1596) retorted that, 'While social scientists are free to appropriate the word to draw a useful distinction in their field, it is not incumbent on the rest of us to do so.' In a subsequent analysis, Pearson (1997) shows that there has been a linear increase in articles that misuse gender for sex. It is a practice that appears to continue and one that perpetuates biology as the source of variation.

Examples of the misapplication of the terms sex and gender are abundant. For instance, Johnson in a 1995 publication, correctly uses the term sex in describing the separation of the X and Y chromosome and then incorrectly substitutes gender a year later in a discussion of 'gender preselection in mammals' (Johnson 1996). Other instances of the substitution of gender for sex abound (Cizza et al. 1996; Hanley et al. 1996; Murata and Masuda 1996; Serrat and de Herreros 1996; Aden et al. 1997; Botchan et al. 1997; Palmer et al. 1997). Even when dealing with social groups where gender may be appropriate, the term is often misused (el-Hazmi *et al.* 1994; Fellous 1997). When Carlin (1996) argues that researchers in other fields have no obligation to accept the distinctions that are useful in other disciplines, he overlooks the power of discriminating between a term that emphasizes biology, with one that makes a social distinction. Even more surprising is Paech's (1996) suggestion that since sex and gender are both social constructs, they are interchangeable. Paech appears to be unaware of what is lost by not making the distinction. Gender is a reflection of what the social system believes to be a biological reality. More importantly, the behavioral practices that reflect gender expectation may have biological outcomes.

There are numerous examples that demonstrate the importance of making a distinction between sex and gender. For example, health practitioners' perceptions that they are dealing with a biological problem rather than a behavioral problem may influence their treatment of the condition. The distinction of sex and gender in prehistory remains an area that has been understudied. Gero and Conkey (1991) provided one of the seminal efforts to recognize the importance of gender in interpreting the archaeological record. The cooperation of biological anthropologists is an essential aspect of uncovering information that allows for an understanding of the importance of gender. Bumsted *et al.* (1990) provide an example of the complexity of unraveling gender differences in nutrition. They have the archaeological context of the population, and a technique to reconstruct diet. Bumsted and her colleagues used stable isotopes to determine differences in diet related to gender. In Chapter 10, Grauer and coworkers note that the most basic issue of gender differences depends on our ability to correctly determine the sex of the skeleton. Without an accurate means for sex determination, the discussion of gender differences is meaningless.

Skeletal biologists who specialize in paleopathology are becoming more sensitive to the issue of sex and gender. There is a plethora of publications that discuss the use of skeletal features for determining the sex of the individual under study. As paleopathologists have begun to understand how behavior affects the risk of pathology, and how gender influences behavior, it has become a more frequent topic of interest. The distinctions between sex and gender are becoming more implicit and explicit in these studies. Paleopathologists are now able to raise more sophisticated issues related to gender and pathology, and to use the data to test hypotheses. It is not surprising that gender continues to come to the forefront as a relevant issue in paleopathology (e.g., Grauer 1991). The pattern of pathology is not a matter of chance but reflects the adaptation of populations. Behavioral differences that are gender-based can affect the pathological profile of a population. In addition, differential access to resources based on gender is a critical factor in producing pathology. One of the difficulties that we have in measuring biological outcomes is how to differentiate results of physiological sex differences from the social aspects of gender. William Stini (1985) discussed this issue in his analysis of the impact of nutritional deficiencies on sexual dimorphism in human stature. From a theoretical perspective, females should be able to resist nutritional deficiencies because of the buffering impact of the hormonal system. Stini (1985) argues that if all things are equal and that males and females are equally subjected to nutritional stress, there should be a greater reduction in the stature of males than females. However, if females are subjected to greater nutritional stress because of differential access to food, then they may suffer a greater percentage reduction in stature than the males (see Storey, Chapter 9, for the application of this hypothesis to an archaeological population). Ortner (Chapter 6) raises a similar issue with respect to the greater immune reactivity in women than in men. He presents empirical evidence to test the hypothesis and offers an evolutionary explanation for the differences. Ortner suggests that gender-related differences in immunity may be related to differential selection because of the women's role in child bearing and nurturing. This would represent a period of increased vulnerability to infection. There is evidence that hormonal differences in males and females affect immunological competence (Sapolsky 1994). In the examination of life tables constructed for prehistoric and ancient populations, women show a pattern of increased mortality during child-bearing years (Green *et al.* 1974; Moore *et al.* 1975). However, women show a decrease in mortality in the later years and experience greater longevity. It is interesting to note that these differences are apparent in most life tables constructed for populations until the beginning of the twentieth century, when they begin to show decreased differences. Recent changes in the lifestyle of women have further reduced these differences.

A comparative method may be used to test hypotheses of gender differences in lifestyle, status, nutrition, and workload. In fact, paleopathology as a science depends on the development of scientific methodologies based on comparative methods. The delayed scientific development of paleopathology is due to the lack of problem-oriented research and a reliance on the newest technology to drive research agendas. Skeletal biologists, using the most advanced medical technology, assume that they are at the forefront of science (Armelagos *et al.* 1982). Substantive research questions are often secondary to the technology applied. The criticism of technology-driven paleopathology should not be interpreted as an argument for rejecting technological advances. Paleopathology would be well served if the new technology were used in conjunction with a methodology that permits the exclusion of alternative hypotheses. Platt (1964) suggests the use of an inductive approach that he calls 'strong inference' as a means for hypothesis testing in science. In spite of the long history of inductive inference, it has not penetrated the methodology of paleopathology; yet it could be an effective means for transforming paleopathology into a true science. While strong inference is most effectively applied to sciences with experimental possibilities, it can be useful in non-experimental sciences, although the application to the latter does require modification. Since there is no possibility of carrying out experiments within the field of paleopathology, the researcher must rely on comparative analysis for 'natural' experiments.

Larsen has used such an approach in his analysis of prehistoric foragers who lived in what is now coastal Georgia and Northern Florida, USA (Larsen and Ruff 1993; Larsen and Harn 1994; Larsen, Chapter 11). In earlier publications, he showed the impact of the shift to agriculture on health and notes the impact of European contact on these populations (Larsen and Milner 1993). In the pre-contact period, females have a higher prevalence of dental caries and periostitis that he believes is related to relative access to maize. After contact, the social, political, and economic changes are so great that it affects both males and females, and sex differences disappear. Larsen also provides some intriguing analysis of bone architecture that suggests that after contact the Europeans were using women as the bearers of burden.

Osteoporosis and osteopenia (Weaver, Chapter 3) illustrate the interplay between sex and gender in an analysis of a problem. Osteoporosis (the loss of bone mass with age) is one of the most serious health problems facing the elderly living in the developed nations. In the United States, 1.5 million women are afflicted with osteoporosis, a condition that increases their risk for fractures of the hip and vertebra. In 1996 it was reported that 300,000 women suffered hip fractures, and the problem will continue to grow as the nation's population ages. Medical and nursing costs have reached 10–20 billion dollars a year in the United States, with projected costs of 240 billion dollars in the next 50 years (Lindsay 1995). From an evolutionary perspective, osteoporosis became a health problem as longevity in humans increased. The increase in life span, with a greater number of individuals reaching these older ages, has created one of the most significant health problems in the world today.

Studies conducted on prehistoric populations, primarily from Sudanese Nubia, North Africa, document the patterning of bone loss and maintenance (Dewey *et al.* 1969a,b). Prehistoric populations living from 10,000 to 1000 years ago experienced two distinct types of bone loss. Many women between their twentieth and thirtieth year lose a significant amount of bone (osteopenia) that appears to be related to the demands of pregnancy, lactation, and a diet that is poor in calcium (Martin *et al.* 1981). The production of milk during lactation extracts calcium from the bone, and in the presence of under-nutrition, this calcium may not be replaced as women grow older (Martin and Armelagos 1985). While these women do not show the clinical problems of bone fractures, it indicates that diet is an important component of bone health in younger women. The examination of children by Van Gerven *et al.* (1985), and Armelagos *et al.* (1982), shows that their bone development and maintenance are also affected. While they show a relatively slight decrease in long bone growth and a significant deficit in cortical wall development, the indications of increased bone resorption and a lack of mineralization are part of the process. The dietary aspect of the problem focuses attention towards gender as a relevant factor in access to resources.

A second pattern of bone loss is related to the aging process. People who lived in the prehistoric period began to lose bone following their thirtieth year. In this pattern of loss, the prehistoric populations are similar to living populations. In both living and extinct populations, males and females experi-ence a decrease in bone mass, but the process is quite different between the sexes. Females, because they have less bone than males as they approach middle age, are especially at risk. In addition, after menopause, the rates of loss may increase because of a decrease in the production of estrogen (a hormone essential for maintaining bone in women). In this instance, hormonal differences related to sex are the focus of attention. There is, however, a major difference in the amount of bone that modern and prehistoric women lose. By 50 years of age, ancient Nubians had lost about 15% of their bone mass, however, they did not suffer from the debilitating fractures that plague modern women. In Nubia, only 4% of the women reached 50 years of age and most died soon after.

Today, women over the age of 50 appear to be at greater risk for bone loss as menopause results in a decrease in estrogen. As more women are living longer they are therefore losing more bone. In the United States, 75% of the population reaches their sixtieth year, 29% their eightieth year, and 6% reach their ninetieth year. It may seem a paradox, but the improvement in living conditions that increases longevity in modern nations has created one of women's most significant health problems.

Much of the research has involved the impact of the subsistence shift to primary food production. For example, the premature loss of bone in women during the reproductive period in prehistoric Nubians and the impact on the growth and development of the children suggest that there is a differential impact of dietary change. The reduction of birth spacing to meet the increase in mortality and the economic importance of children that characterize Neolithic populations suggest that women and children were bearing the cost of this transformation. It is a pattern that is played out in many of the peasant populations that live in the Third World. Weaver (Chapter 3) is correct in stating that osteopenia is not a problem that affects only women. Martin *et al.* (1987) and Rose (1985) showed in their studies of post emancipation populations from Ceder Grove Arkansas, USA that both women and men were seriously affected by premature bone loss. In this case, the stresses associated with the adaptation of these populations had a significant affect on both sexes.

Given the importance of distinguishing between sex and gender in prehistory, it is not surprising that the distinction will be useful in a contemporary setting. There is, for example, a concern for the occurrence of osteoporosis (Anonymous 1996) and breast cancer in men (Seeman 1995; Memon and Donohue 1997). The analysis of the archaeological record provides a means for examining differences in sex and gender from an evolutionary perspective. Changes in men's health (Sabo and Gordon 1995) and women's life expectancy (Williamson and Boehmer 1997) can be investigated. Recently, it has been asserted that mortality differences in rheumatoid arthritis have a gender and age component (Anderson 1996). Differences in depression (Compas *et al.* 1997) must also consider the issue of gender. In education (Field and Lennox 1996; Zelek *et al.* 1997), training (Wilson and Boulter 1997), and practice (Wiggins 1996), gender has become an issue for health care professions. The role of men in nursing (Evans 1997), gender in doctor–patient relationships (Greatrex 1997; Kerssens *et al.* 1997), and recruiting women physicians for specialties such as gastroenterology (Wolf 1997) have entered the debate. The role of gender in international health (Sargent and Brettell 1996) has surfaced as one of the most critical issues of this decade. Women's access to health care facilities in India (Buckshee 1997; Roberts, Chapter 7), violence against children in Barbados (Handwerker 1996), excess female mortality in Somalia (Aden *et al.* 1997), and access to nutritional resources (Backstrand *et al.* 1997) have brought gender to the forefront of international health issues. International health care providers are now beginning to consider the issues raised by the distinction between sex and gender (Wijeyaratne 1994; Pfannenschmidt *et al.* 1997). Anthropology, with its biocultural perspective, initiated and championed the distinction between sex and gender. The importance of discriminating between these concepts and their application to problems in contemporary and prehistoric populations is now well established in anthropology. The trend toward distinguishing between sex and gender in the fields of medicine and international public health suggests the importance of making the distinction in other disciplines. It is now incumbent for others to follow this advancement.

References

Aden AS, Omar MM, Omar HM, Hogberg U, Persson LA, and Wall S (1997) Excess female mortality in rural Somalia – is inequality in the household a risk factor? *Social Science and Medicine* **44**(5):709–15.

Anderson ST (1996) Mortality in rheumatoid arthritis: do age and gender make a difference? *Seminars in Arthritis and Rheumatism* **25**(5):291–6.

Anonymous (1996) Looking at gender differences. *Tecnologica* **1**:3–6.

Armelagos GJ, Carlson DS, and Van Gerven DP (1982) The theoretical foundation of development of skeletal biology. In F Spencer (ed.), *A History of Physical Anthropology, 1930–1980.* Academic Press, pp. 305–28.

Backstrand JR, Allen LH, Pelto GH, and Chavez A (1997) Examining the gender gap in nutrition: an example from rural Mexico. *Social Science and Medicine* **44**(11):1751–9.

Botchan A, Hauser R, Gamzu R, Yogev L, Paz G, and Yavetz H (1997) Sperm separation for gender preference: methods and efficacy. *Journal of Andrology* **18**(2):107–8.

Buckshee K (1997) Impact of roles of women on health in India. *International Journal of Gynecology & Obstetrics* **58**(1):35–42.

Bumsted MP, Brooker J, Barnes R, Boutton T, Armelagos GJ, Lerman JC, and Brendel K (1990) Recognizing women in the archeological record. In SM Nelson and AB Kehoe (eds.), *Powers of Observation: Alternative Views in Archeology*, Volume 2. *Archeological Papers of the American Anthropological Association*, pp. 89–101.

Carlin NF (1996) Sex and gender. *Science* **274**(5293):1595-6.

Cizza G, Brady LS, Esclapes M, Blackman MR, Gold PW, and Chrouso GP (1996) Age and gender influence basal and stress-modulated hypothalamic–pituitary–thyroidal function in Fischer 344/N rats. *Neuroendocrinology* **64**(6):440–8.

Compas BE, Oppedisano G, Connor JK, Gerhardt CA, Hinden BR, Achenbach TM, and Hammen C (1997) Gender differences in depressive symptoms in adolescence: comparison of national samples of clinically referred and nonreferred youths. *Journal of Consulting and Clinical Psychology* **65**(4):617–26.

Dewey JR, Armelagos GJ, and Bartley MH (1969a) Femoral cortical involution in three archaeological populations. *Human Biology* **41**:3–28.

Dewey JR, Bartley MH, and Armelagos GJ (1969b) Rates of femoral cortical bone loss in two Nubian populations utilizing both normalized and non-normalized data. *Clinical Orthopaedics* **65**:61–6.

el-Hazmi MA, Warsy AS, Addar MH, and Babae Z (1994) Fetal haemoglobin level – effect of gender, age and haemoglobin disorders. *Molecular and Cellular Biochemistry* **135**(2):181–6.

Evans J (1997) Men in nursing: issues of gender segregation and hidden advantage. *Journal of Advanced Nursing* **26**(2):226-31.

Fellous JM (1997) Gender discrimination and prediction on the basis of facial metric information. *Vision Research* **37**(14):1961–73.

Field D and Lennox A (1996) Gender in medicine: the views of first- and fifth-year medical students. *Medical Education* **30**(4):246–52.

Gero JM and Conkey MW (Eds.) (1991) *Engendering Archaeology: Women in Prehistory*. Oxford: Blackwell.

Grauer AL (1991) Life Patterns of Women from Medieval York. In D Walde and ND Willows (eds.), *The Archaeology of Gender*. Calgary, Canada: Chocmool Archaeological Society, University of Calgary, pp. 407–13.

Greatrex TS (1997) Effects of gender on the doctor–patient relationship. *MD Computing* **14**(4):266–73.

Green S, Green S, and Armelagos GJ (1974) Settlement and mortality of the Christain Site (1050 AD–1300 AD) of Meinarti (Sudan). *Journal of Human Evolution* **3**:297–316.

Handwerker WP (1996) Power and gender: violence and affection experienced by children in Barbados, West Indies. *Medical Anthropology* **17**(2):101–28.

Hanley K, Rassner U, Jiang Y, Vansomphone D, Crumrine D, Komuves L, Elias PM, Feingold KR, and Williams ML (1996) Hormonal basis for the gender difference in epidermal barrier formation in the fetal rat. Acceleration by estrogen and delay by testosterone. *Journal of Clinical Investigation* **97**(11):2576–84.

Johnson LA (1995) Sex preselection by flow cytometric separation of X and Y chromosome-bearing sperm based on DNA difference: a review. *Reproduction, Fertility, and Development* **7**(4):893-903.

Johnson LA (1996) Gender preselection in mammals: an overview. *Deutsche Tierarztliche Wochenschrift* **103**(8–9):288–91.

Kerssens JJ, Bensing JM, and Andela MG (1997) Patient preference for genders of health professionals. *Social Science and Medicine* **44**(10):1531–40.

Larsen CS and Harn DE (1994) Health in transition: disease and nutrition in the Georgia Bight. In KD Sobolik (ed.), *Paleonutrition: The Diet and Health of Prehistoric Americans*. Center for Archaeological Investigations, Occasional Paper Number 22. Edwardsville, Illinois: Southern Illinois University at Carbondale, pp. 222–34.

Larsen CS and Milner GR (Eds.) (1993) *In the Wake of Contact: Biological Responses to Conquest*. New York: Wiley-Liss.

Larsen CS and Ruff CB (1993) The stresses of conquest in Spanish Florida: structural adaptation and change before and after contact. In CS Larsen and GR Milner (eds.), *In the Wake of Contact: Biological Responses to Conquest*. New York: Wiley-Liss, pp. 21–34.

Lindsay R (1995) The burden of osteoporosis: cost. *American Journal of Medicine*. **98**(2A):9S–11S.

Martin DL and Armelagos GJ (1985) Skeletal remodeling and mineralization as indicators of health: an example from prehistoric Sudanese Nubia. *Journal of Human Evolution* **14**:527–37.

Martin DL, Armelagos GJ, Mielke JH, and Miendl R (1981) Bone loss and dietary stress in prehistoric populations from Sudanese Nubia. *Bulletins et Memoires de la Societe d'Anthropologie de Paris* Tome 8, series XIII(3):307–19.

Martin DL, Magennis AL, and Rose JC (1987) Cortical Bone Maintenance in an Historic Afro-American Cemetery Sample from Cedar Grove, Arkansas. *American Journal of Physical Anthropology* **74**(2):255–64.

Memon MA and Donohue JH (1997) Male breast cancer. *British Journal of Surgery* **84**(4):433–5.

Moore JA, Swedlund AC, and Armelagos JG (1975) The use of life table in paleodemography. Population studies in archaeology and biological anthropology. Memoir 30. *American Antiquity* **40**(2):57–70.

Murata K and Masuda R (1996) Gender determination of the Linne's two-toed sloth (*Choloepus didactylus*) using SRY amplified from hair. *Journal of Veterinary Medical Science* **58**(12):1157–9.

Paech M (1996) Sex or gender? A feminist debate for nurses. *Contemporary Nurse* **5**(4):149–56.

Palmer LJ, Pare PD, Faux JA, Moffatt MF, Daniels SE, Le Souef PN, Bremner PR, Mockford E, Gracey M, Spargo R, Musk AW, and Cookson WOCM (1997) Fc epsilon R1-beta polymorphism and total serum IgE levels in endemically parasitized Australian aborigines. *American Journal of Human Genetics* **61**(1):182–8.

Pearson GA (1996) Of sex and gender. *Science* **274**(5285):328–9.

Pearson GA (1997) Engendering confusion: misuse of the terms 'sex' and 'gender' in scientific literature. http:/www.albion.edu.fac/biol/pearson.

Pfannenschmidt S, McKay A, McNeill E, and Family Health International (1997) *Through a Gender Lens: Resources for Population, Health and Nutrition Projects*. Gender Working Group. USAID.

Platt JR (1964) Strong inference. *Science* **146**(3642):347–53.

Rose JC (Ed.) (1985) *Gone to a Better Land: A Biohistory of a Rural Black Cemetery in the Post-Reconstruction South*. Fayetteville, Arkansas: Arkansas Archeological Survey Research Series Number 25, W. Fredrick Limp, Series Editor.

Sabo DF and Gordon D (1995) *Men's Health and Illness: Gender, Power, and the Body*. Thousand Oaks, California: Sage Publications.

Sapolsky RM (1994) *Why Zebras Don't Get Ulcers*. New York: W. H. Freeman Company.

Sargent CF and Brettell C (1996) *Gender and Health: An International Perspective*. Upper Saddle River, New Jersey: Prentice Hall.

Seeman E (1995) The dilemma of osteoporosis in men. *American Journal of Medicine* **98**(2A):76S–88S.

Serrat A and de Herreros AG (1996) Gender verification in sports by PCR amplification of SRY and DYZ1 Y chromosome specific sequences: presence of DYZ1 repeat in female athletes. *British Journal of Sports Medicine* **30**(4):310–12.

Stini WA (1985) Growth rates and sexual dimorphism in evolutionary perspective. In RI Gilbert, Jr. and JH Mielke (eds.), *The Analysis of Prehistoric Diets*. Orlando, Florida: Academic Press, pp. 191–226.

Van Gerven DP, Hummart JR, and Burr DB (1985) Cortical bone maintenance and geometry of the tibia in prehistoric children from Nubia's Batn el Hajar. *American Journal of Physical Anthropology* **66**(3):275–80.

Wiggins C (1996) Counting gender: does gender count? *Journal of Health Administration Education* **14**(3):379–88.

Wijeyaratne P (1994) Gender, health, and sustainable development: a Latin American perspective. Proceedings of a workshop held in Montevideo, Uruguay, 26–29 April 1994. Ottawa: International Development Research Centre, pp. 263.

Williamson JB and Boehmer U (1997) Female life expectancy, gender stratification, health status, and level of economic development: a cross-national study of less developed countries. *Social Science and Medicine* **45**(2):305–17.

Wilson JA and Boulter PS (1997) Targeting medical students to promote women in
    surgery. *Journal of the Royal College of Surgeons of Edinburgh*
    42(4):217–18.
Wolf JL (1997) Gender equality in gastroenterology: an achievable goal?
    Gastroenterology **113**(2):684–6.
Zelek B, Phillips SP, and Lefebvre Y (1997) Gender sensitivity in medical
    curricula. *Canadian Medical Association Journal* **156**(9):1297–300.

2

Sex-related patterns of trauma in humans and African apes

ROBERT JURMAIN and LYNN KILGORE

Skeletal evidence of traumatic lesions often provides an intriguing interpretive window into the lives of individuals. Traumatic lesions often record severe injuries that can incapacitate individuals or even end their lives. Causes of trauma vary, ranging from the purely accidental to the deliberately aggressive. Whatever the cause, these events, which can compromise an individual's ability to exploit resources or to find mates, can have adaptive consequences when premature death impacts reproductive success.

Comparisons between populations and across species can provide insights regarding adoptive patterns that may be sex specific. Most notably, certain types of injuries are known to show systemic gender-related incidence in a variety of contemporary human populations (Breiting *et al.* 1989; Matthew *et al.* 1996). Are such apparent sex biases in trauma also evident in earlier human populations and among African ape species? Are there broader sex-related adaptive considerations that help explain patterns of injuries in humans as well as other large-bodied hominoids? In this chapter we seek to address these questions.

In the skeletal record, the most common and usually the most obvious manifestations of trauma are healed fractures. In this study we report the incidence of healed traumatic lesions (including those resulting from weapon wounds) in the crania and postcrania (long bones) from two archaeological samples of humans and from samples of the three species of African great ape (*Pan troglodytes troglodytes, P. paniscus, Gorilla gorilla gorilla*). The need to evaluate *both* cranial as well as postcranial involvement has recently been emphasized by Berger and Trinkaus (1995), who argue in their analysis of fractures by body segment, that such data can be used to infer the cause of the injuries.

An epidemiological perspective must also be applied to gain a fuller understanding of sex-related injuries in humans and other hominoids. It is necessary to compare adequate samples of several different species to determine whether any general patterns are present. Obviously, humans differ significantly from apes in a number of ways, including mode of locomotion and degree of arboreality, and these differences will no doubt influence postcranial trauma. Indeed, Lovell (1990) has demonstrated some differences in frequency of postcranial trauma among great ape species, and has attributed the differences to variation in the degree of arboreality. However, the causes of cranio-facial injury *may* be more similar across hominoid species. Most especially, the influence of inter-individual aggression could potentially produce inter-species similarities in cranio-facial patterns of trauma. Interpretations of these similarities are intriguing, but of course, not without controversy.

Studies of trauma in human archaeological samples have also adopted an epidemiological approach, particularly research into long bone injuries (Elliott-Smith and Wood Jones 1910; Lovejoy and Heiple 1981; Bennike 1985; Jurmain 1991; Grauer and Roberts 1996; Smith 1996; Kilgore *et al.* 1997). A variety of studies have also utilized a similar approach in analyzing cranial trauma (Newman 1957; Tyson 1977; Bennike 1985; Walker 1989; Owsley *et al.* 1994; Webb 1995; Jurmain and Bellifemine 1997). Little systematic epidemiological work has been done on great apes, but prevalence rates have been reported by Lovell (1990) and Jurmain (1997).

Weapon wounds are commonly reported in Old World human populations (e.g., Wells 1964; Bennike 1985; Hawkes 1989; Waldron 1994). Instances in the New World (mostly involving projectile injury), have been reported by Pfeiffer (1985), Bridges (1996), and Smith (1996), with the highest incidences documented in prehistoric California (Tenney 1986; Lambert and Walker 1991; Jurmain 1991; Lambert 1994; Jurmain and Bellifemine 1997). Analogous injuries in nonhuman primates (bite wounds) have not, until recently, received much attention from skeletal biologists (Jurmain 1997).

To date, few analyses have focused on sex-related trauma in humans. Some exceptions include investigations by Merbs (1983), Walker (1989), Wilkinson and Van Wagenen (1993), Webb (1995), Smith (1996), Bridges (1996), and Grauer and Roberts (1996). In some studies, sex-related patterns received little attention, as few differences in traumatic involvement were found between the sexes (e.g., Kilgore *et al.* 1997). Information on sex-related trauma in nonhuman primates is virtually non-existent.

Table 2.1. *Skeletal samples utilized*

	N	Males	Females
Human samples			
Kulubnarti			
long bones	147	67	80
crania	131	61	70
Ala-329			
long bones	248	138	110
crania	238	134	104
Total			
long bones	395	205	190
crania	369	195	174
Great ape samples			
Chimpanzees			
long bones	92	28	64
crania	116	27	89
Gorillas			
long bones	62	28	34
crania	135	69	66
Bonobos			
long bones	15	*	*
crania	56	29	27
Total			
long bones	169	56	98
crania	307	125	182

* undetermined
N: number of sexed adult individuals with specified elements

Materials and methods

The human samples utilized in this study were derived from two archaeological sites, one in North Africa (Sudanese Nubia), and the other from central California, USA. In total, 395 adult human postcranial skeletons (with long bones) that could be assigned an age and sex, and 369 crania were surveyed. Samples of the three species of African apes were also investigated. Great ape material analyzed included a total of 169 sexed adult postcrania with long bones and 307 crania (Table 2.1).

The Sudanese Nubian remains were excavated from two temporally-overlapping cemeteries at Kulubnarti, dating from approximately the sixth to sixteenth centuries AD (Van Gerven *et al.* 1981;1995). Preservation was generally excellent. In fact, many individuals were partially mummified. The central California skeletal remains derive from site Ala-329 located on the east side of San Francisco Bay. The site was occupied for over 2000 years, from 500 BC to AD 1500. Again, preservation was good as the remains were interred in a shell/earthmound (Jurmain 1990; Leventhal 1993; Jurmain and Bellifemine 1997).

The great ape materials comprised portions of two 'wild shot' museum collections. The chimpanzee (*Pan troglodytes troglodytes*) and lowland gorilla (*Gorilla gorilla gorilla*) skeletal materials are curated at the Powell-Cotton Museum, Birchington, Kent, UK. These animals were collected primarily during the 1930s, from the Southern Cameroons. The bonobo (*P. paniscus*) sample, collected during much of the twentieth century in Zaire, is curated at the Musee Royal de l'Afrique Centrale in Tervuren, Belgium. Except for an occasional missing element, most pongid postcranial skeletons were complete. However, owing to collection biases, all three species had considerably more crania than postcrania available for examination.

In both the human and nonhuman skeletal material, diagnosis of fractures was based on macroscopic observations. These included the presence of callus formation and/or angular deformation in the facial and long bones, and depression fractures of the cranial vault. In most cases, observation was supplemented by radiographic examination (except for the bonobo materials, as radiography was not available in Tervuren).

Results

Long bone fractures

At Kulubnarti, 3.7% ($n=67$) of 1788 long bones from all adult individuals displayed healed fractures (Table 2.2). Of 147 sexed individuals with long bones, 32.7% ($n=48$) displayed fractures and of the 48 with long bone fractures, 43.8% ($n=21$) were male and 56.2% ($n=27$) were female. That is, both sexes displayed similar frequency patterns. It was also interesting to note that of the 48 individuals with fractures, 27.1% ($n=13$) had more than one fracture. Hence, a surprisingly high percentage (8.8%) of all sexed individuals with long bones had multiple fracture involvement (defined here as two or more lesions). Of the 13 individuals with two or more fractures, 10 (76.9%) were female.

More precise patterns can be revealed through an analysis of involvement by skeletal element (Table 2.3). With the exception of the radius, the incidence of healed fractures was very similar in both sexes for all elements. For the radius, however, females showed a markedly higher incidence of healed fractures, exceeding males by 650% ($p < 0.01$).

Table 2.2. *Patterns of long bone fractures*

	Total long bones of sexed adults			Total sexed adults with long bones			Males with long bones			Females with long bones		
	N	n	(%)	N	n	(%)	N	n	(%)	N	n	(%)
Kulubnarti	1788	67	(3.7)	147	48	(32.7)	67	21	(31.3)	80	27	(33.8)
Ala-329	1953	36	(1.8)	248	23	(9.3)	138	12	(8.7)	110	11	(10.0)
Chimpanzees	1267	22	(1.7)	92	20	(21.7)	28	9	(32.1)	64	11	(17.2)
Gorillas	879	13	(1.5)	62	11	(17.7)	28	3	(10.7)	34	8	(23.5)
Bonobos	197	2	(1.0)	**	**		**	**		**	**	

** Sample too small for analysis
N: Total sample size with long bones
n: Total sample with healed long bone fractures
%: Percentage of total sample with long bones

Table 2.3. *Prevalence of healed long bone fractures by skeletal element at Kulubnarti, Nubia*

	Males			Females		
	N	n	(%)	N	n	(%)
Clavicle	124	1	(0.8)	138	0	(0.0)
Humerus	128	4	(3.1)	148	6	(4.0)
Radius	124	2	(1.6)	135	14	(10.4)
Ulna	123	16	(13.0)	137	18	(13.1)
Femur	130	1	(0.7)	151	3	(2.0)
Tibia	117	1	(0.8)	115	1	(0.9)
Fibula	116	1	(0.9)	102	2	(2.0)
Total	862	26	(3.0)	926	44	(4.8)

N: Number of bones available for examination
n: Number of bones with healed fractures
%: Percentage of total number of elements with healed fractures

The California Indian sample from Ala-329 showed consistently fewer long bone fractures than were seen in the Kulubnarti materials. Out of a total of 1953 complete long bones from adult individuals, 1.8% (*n*=36) displayed fractures (Table 2.2). Of the 248 adults who could be sexed reliably, 9.3% (*n*=23) had at least one fracture. Of those individuals with fractures, 52.2% (*n*=12) were male (comprising 8.7% of all males with long bones and 4.8% of all sexed adults), and 47.8% (*n*=11) were female (comprising 10.0% of all females with long bones and 4.4% of all sexed adults). These rates of

Table 2.4. *Prevalence of long bone healed fractures by skeletal element at Ala-329, Central California*

	Males			Females	
	N	n	(%)	N	n (%)
Clavicle	138	2	(1.4)	145	0 (0.0)
Humerus	134	1	(0.7)	155	0 (0.0)
Radius	138	6	(4.3)	148	7 (4.7)
Ulna	130	7	(5.4)	142	7 (4.9)
Femur	140	0	(0.0)	155	0 (0.0)
Tibia	140	2	(1.4)	151	1 (0.7)
Fibula	123	1	(0.8)	114	0 (0.0)
Total	943	19	(2.0)	1010	15 (1.5)

N: Number of long bones available for examination
n: Number of long bones with healed fractures
%: Percentage of total number of long bones available for exam

involvement by sex were essentially identical. Of the 10 individuals with multiple fractures, 6 were male and 4 were female. This difference was also not statistically significant. The distribution of healed long bone fractures by skeletal element (Table 2.4) likewise did not reveal any sex-related patterns, or statistically significant differences between skeletal elements.

The prevalence of long bone fractures in the three great ape species appears similar to that observed in the Ala-329 sample (although the long bone sample of bonobos from a total of 15 individuals was too limited to permit definitive conclusions) (Table 2.2). Among the chimpanzees, 1.7% ($n=22$) of the 1267 available long bones displayed fractures, and among the gorillas 1.5% ($n=13$) of the 879 long bones were affected. Of the 197 bonobo long bones available for analysis, 1.0% ($n=2$) displayed fractures. Out of 92 chimpanzees, 21.7% ($n=20$) individuals had at least one fracture. Nine of the individuals were male, 11 were female. Although 32.1% of all males and only 17.2% of all females displayed fractures, the difference was not statistically significant ($p=0.18$). Of the 62 sexed gorillas, 17.7% ($n=11$) showed at least one long bone fracture. Of these, 27.3% ($n=3$) were male (comprising 10.7% of all males in the sample), and 72.7% ($n=8$) were female (comprising 23.5% of all females). Again, the difference was not statistically significant ($p=0.33$).

Cranial Trauma

Sex-related patterns of trauma are considerably more apparent in the cranium than in the postcranial skeleton. At Ala-329, of the 238 adult crania of individuals with ascertainable sex, four (all males) showed depression fractures of the cranial vault (Figure 2.1). In addition, two other males displayed facial fractures (of the zygomatic and zygomatic arch respectively). The prevalence of cranio-facial injury in the Ala-329 sample is thus 6/134 (4.5%) in males and 0/104 in females, with the difference bordering on statistical significance (p=0.07). One other cranial fracture was present, in an adolescent (aged 13–15 years) from Ala-329, provisionally ascertained to be male when sexed by pelvic morphology.

In the Kulubnarti materials, a much lower prevalence of cranial injury was found. Only one lesion (a small depression fracture of the frontal bone) was observed, in a young male. Of additional interest, this individual displayed a variety of skeletal features most likely diagnostic of pituitary dwarfism.

In our analysis of cranial trauma in the great apes samples, six cranial vault injuries and 10 facial fractures were observed in the combined samples for the three species (see Table 2.5). A marked difference in sex distribution of these lesions was observed. While all six vault injuries were found in females, 9 out of 10 facial lesions were found in males. This variation in anatomical

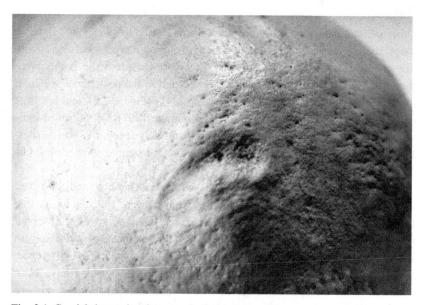

Fig. 2.1 Cranial depression fracture, from Ala-329; adult male.

Table 2.5. *Sex-related patterns of cranial trauma*

	N	Vault fractures			Facial fractures			Bite wounds		
		Total	*M*	*F*	*Total*	*M*	*F*	*Total*	*M*	*F*
Human samples										
Kulubnarti	131	1	1	0	0	0	0	—	—	—
Ala-329	238	4	4	0	2	2	0	—	—	—
Total	369	5	5	0	2	2	0	—	—	—
Great ape samples										
Chimpanzees	116	1	0	1	3	2	1	3	3	0
Gorillas	135	5	0	5	5	5	0	5	5	0
Bonobos	56	0	0	0	2	2	0	0	0	0
Total	307	6	0	6	10	9	1	8	8	0

N: Number of sexed adults with crania

location of cranio-facial injuries was highly significant ($p=0.001$). Considering all instances of cranio-facial trauma (including bite wounds), males displayed a total of 17 lesions, and females a total of seven; the difference was statistically significant ($p=0.007$).

Wounds

Other skeletal indicators of trauma are wounds caused by 'weapons'. In some cases, such lesions provide unambiguous evidence of interpersonal aggression. For example, in the Ala-329 sample, 12 embedded projectile points were found in 10 individuals. Of these, 11 were found in the postcranial skeleton, mostly in the vertebrae or pelvis (Jurmain 1991) (Figure 2.2), and one was embedded in a cranium (Jurmain and Bellifemine 1997). Of the seven individuals with projectile wounds who could be reliably sexed, five were males. In contrast to the California group, the Kulubnarti sample showed no indications of weapon injury.

Among apes, there is a strong corollary with the weapon wounds observed in human groups. In apes, however, the wounds were inflicted by canine teeth. In the combined ape sample, eight healed bite wounds were observed (Table 2.5 and Figure 2.3); three of these were in chimpanzees and five were seen in gorillas. All eight of these injuries were found in males (sex difference, $p=0.002$). It should also be noted that two other bite wounds have been observed in the chimpanzee skeletons from Gombe National Park, and here too, both lesions were seen on males (Jurmain 1989,1997).

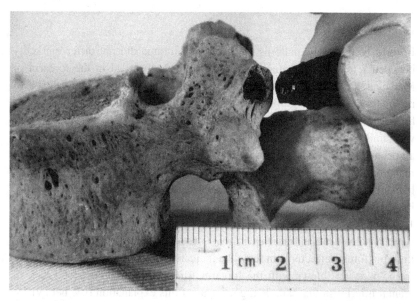

Fig. 2.2 Embedded obsidian projectile fragment, Ala-329; adult male. The remaining portion of the obsidian point was recovered with the body and, ante-mortem, was probably contained within soft tissue.

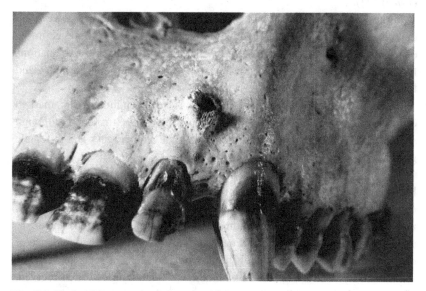

Fig. 2.3 Healed bite wound; adult male chimpanzee.

Discussion

In most respects, the patterns of postcranial trauma did not differ markedly between females and males. An exception was observed in the forearm of Medieval Sudanese Nubians from Kulubnarti, for which the prevalence of radial fractures was significantly higher in females. Women also showed a higher rate of multiple fractures. Although the prevalence of ulnar lesions was similar in men and women, in women, fracture of the ulna was frequently associated with radial fracture. This difference in pattern was especially marked in the later cemetery at Kulubnarti (dated *c*. AD 1000–1600). Drawing behavioral inferences from such skeletal patterning is tempting, but should be tempered by an inability to test the interpretations (Kilgore *et al.* 1997; Jurmain, unpublished data). For example, it might be argued that the more typical isolated ulnar fractures in males (contrasted with the more frequent multiple forearm fractures in females) indicates such injuries in males resulted from 'parrying' blows with the forearm. However, the almost total lack of cranial injuries in this sample argues against this explanation (see below). At present, the most reasonable explanation for the majority of postcranial fractures at Kulubnarti is that they resulted from falls. This explanation is entirely possible, given the extremely rough terrain found in this part of Nubia (Kilgore *et al.* 1997).

In the Ala-329 sample, the prevalence of postcranial injury is lower than that observed at Kulubnarti. There is no indication of a sex difference in the distribution of injuries. Again, accidental falls are reasonable explanations for these injuries, with falls being uniformly distributed between males and females.

Postcranial trauma among African apes was also uncommon. Again, marked difference in prevalence between females and males was not found (although such injuries were more frequently found in male chimpanzees and in female gorillas).

The only dramatic sex-related differences in the patterns of traumatic lesions were observed in the crania. Clearly, not all groups (or species) exhibited equal risk of cranial injuries. Indeed, especially for facial involvement (including those lesions relating to weapon use), populations and sexes apparently varied greatly in risk. For example, at Kulubnarti the incidence of cranio-facial injury was extremely low, with involvement seen in only one individual. By contrast, at Ala-329, numerous cranial injuries were observed, affecting both the cranial vault and face, and, in this group such lesions were confined to males. The difference in prevalence of cranial injury between these two populations was statistically significant ($p=0.03$).

In the African ape samples, cranio-facial injury displayed a distinctive sex-related pattern. While vault injuries were found only in females, facial injuries were dramatically more common in males. Combining facial fractures with bite wounds, a total of 18 injuries were observed in the combined pongid samples. Of these, 17 (94.4%) were found in males ($p < 0.0001$) (see Table 2.5).

Behavioral interpretations must be pursued with caution. Moreover, overall patterns of traumatic involvement are the best basis from which to draw conclusions (Berger and Trinkaus 1995; Jurmain, unpublished data). To use this perspective, the relative incidence of trauma in different anatomical segments should be assessed. Specifically, the relative involvement of the cranium and the postcranium must be examined. Smith (1996) and Kilgore *et al.* (1997) have made the argument that if interpersonal aggression was common in an ancient society (or in hominoid remains), it should be reflected in high cranio-facial involvement. Thus, in human skeletal remains from Tennessee, USA (Smith 1996), as well as from Nubia (Kilgore *et al.* 1997), the absence of cranio-facial involvement suggests interpersonal aggression was most likely rare. It is also possible that some postcranial lesions, frequently identified as 'parry' fractures, were *not* related to interpersonal violence. Apparently, in these cultures *neither* males nor females were at much risk of such injury.

What about groups in which interpersonal aggression was more common? It is important to be able to identify the contexts in which the aggression has occurred most frequently. In some circumstances, as with direct evidence of weapon use, the interpretation is quite clear. Evidence of projectile weapons has been clearly demonstrated at several prehistoric locales in California (Tenney 1986; Lambert and Walker 1991; Jurmain 1991; Lambert 1994).

Violence in the form of mutilation and/or scalping, has also been inferred from skeletal remains, particularly in the American Midwest (Owsley *et al.* 1977; Willey 1990; Milner *et al.* 1991; McNamara 1996; Grauer 1995) and Southeast (Bridges 1996). Often, definitive skeletal signatures of deliberate trauma are associated with high prevalence of cranio-facial fractures. The least diagnostic lesions from which to interpret aggression are fractures in the postcranial skeleton (e.g., 'parry' fractures).

Patterns of facial injuries seen in great apes are also potentially informative. However, the distribution of cranial vault injuries among gorillas and chimpanzees (restricted to females) is puzzling, particularly because they are largely independent of facial injuries.

If we assume that skeletal indicators of aggression are reliable, the sex-related patterns suggest some intriguing behavioral trends. In the human sample from California reviewed here, where evidence of aggression appears quite clear, the majority of injuries was found in males. Walker (1989) also observed higher prevalence of cranial injuries among males from southern California. Indeed, in many investigations of populations where popultion controls were employed, males display more cranial injuries than females. There are, however, some notable exceptions where females were equally, if not more frequently, affected than males (Wilkinson and Van Wagenen 1993; Webb 1995). The markedly higher incidence of facial injuries seen in male African apes is also suggestive of aggression, and is supported by ethological documentation of male aggression in these primate groups (e.g., Manson and Wrangham 1991).

The human patterns of trauma are more variable than those seen among great apes. Nevertheless, some contemporary clinical data suggest that strong sex-related trends exist. In a hospital sample in Papua New Guinea, males accounted for 67% of all trauma cases, and a large proportion of these (55%) were the result of aggression (Matthew *et al.* 1996). In instances of head injury, males accounted for 78% of the cases. Similarly, in a Danish study (Breiting *et al.* 1989), males were victims in 72% of the cases. Likewise, in Sweden, males were the victims in 82% and 75% of cases from rural and urban samples respectively (Strom *et al.* 1992). Males in a Chilean sample accounted for 84% of the cases (Aalund *et al.* 1990), and in Argentina, 70% of victims were males (Danielsen *et al.* 1989). In a controlled study of assault victims, conducted in South Africa, 84% were male (Butchart and Brown 1991). Likewise, in a sample of South African children and adolescents, 70% of maxillo-facial injuries occurred in males (Bamjee *et al.* 1996). Data from the United Kingdom have shown that males accounted for 83% of facial fractures in a Scottish sample (Brook and Wood 1983), 84% of facial fractures in a sample from Bristol (Shepherd *et al.* 1990), and 79% of mandibular fractures in a sample from Manchester (Asadi and Asadi 1996). Finally, in a study of repeat assault victims in San Diego, USA, 87% were males (Hedges *et al.* 1995). Even more conclusively, numerous studies suggest that assailants are most frequently male. In Sweden (Strom *et al.* 1992), 91% of assailants were male, and in Denmark (Breiting *et al.* 1989), 95% were male.

The sex-related distribution of cranial trauma is important, especially since it appears consistently among the four closely related species examined in this study. The frequency of male–male aggression in the three African great apes species, as well as in *Homo sapiens*, might thus have an evolutionary basis, with phylogenetic roots perhaps reaching back to the late Miocene.

Human behavior, especially as modulated by culture, displays a wider range of expression than that seen in any apes. This variability can obscure basic patterning, even blinding those already disinclined to search for it. However, the basic neurology and endocrinology of all African hominoids is highly similar. Why, therefore, should we not expect certain evolutionary influences on behavior to be present among these closely-related species? Many anthropologists would disagree with this suggestion, preferring instead to dismiss nonhuman primate data as 'irrelevant'. Rather, they would emphasize the variability of human behavior (and its skeletal expression). This ready willingness to discount evolutionary/adaptive influences has been critiqued elsewhere (Cartmill 1990). For physical anthropologists, in particular, a headlong retreat from evolutionary considerations seems to be misplaced. The patterns of trauma presented here strongly suggest that males and females frequently face different challenges and risks. Moreover, some of these patterns appear to show considerable commonalties across contemporary animal species – including humans.

Acknowledgments

We would like to express our grateful appreciation to the following for their assistance during various stages of this research: Derek Howlett and Malcomb Harman at the Powell-Cotton Museum, and Wim van Neer and Guy Teugels at the Musee Royal de l'Afrique Centrale. We also give our thanks to Dennis Van Gerven at the University of Colorado, Anton Musladen, Alan Leventhal, Viviana Bellifemine, San Jose State University, and the Photography Laboratory, Instructional Resource Center at San Jose State University. We also thank Harry and Betty Early of Birchington for their hospitality, and the Veterinary Hospital, Margate, UK for providing X-ray facilities. Lastly, we are grateful to M.E. Morbeck for her assistance and encouragement with the great ape research. This study was supported, in part, by the L.S.B. Leakey Foundation and the College of Social Sciences, San Jose State University.

References

Aalund O, Danielsen L, and Sanhueza RO (1990) Injuries due to deliberate violence in Chile. *Forensic Science International* **46**:189–202.

Asadi SG and Asadi Z (1996) Site of the mandible prone to trauma: A two year retrospective study. *International Dental Journal* **46**:171–3.

Bamjee Y, Lownie JF, Cleaton-Jone PE, and Lownie MA (1996) Maxillofacial injuries in a group of South Africans under 18 years of age. *British Journal of Oral and Maxillofacial Surgery* **34**:298–302.

Bennike P (1985) *Palaeopathology of Danish Skeletons*. Copenhagen: Akademisk Forlag.

Berger TD and Trinkaus E (1995) Patterns of trauma among the Neandertals. *Journal of Archaeological Science* **22**:841–52.

Breiting VB, Aalund O, Albreksten SB, Danielsen L, Helwig-Larsen K, Jacobsen J, Kjaerulff H, Staugaard H, and Thomsen JL (1989) Injuries due to deliberate violence in areas of Denmark. I. The extent of violence. *Forensic Science International* **40**:183–99.

Bridges P (1996) Warfare and morality at Koger's Island, Alabama. *International Journal of Osteoarchaeology* **6**:66–75.

Brook IM and Wood N (1983) Aetiology and incidence of facial fractures in adults. *International Journal of Oral Surgery* **12**:293–8.

Butchart A and Brown DSO (1991) Non-fatal injuries due to interpersonal violence in Johannesburg – Soweto: incidence, determinants, and consequences. *Forensic Science International* **52**:35–51.

Cartmill M (1990) Human uniqueness and theoretical content in paleoanthropology. *International Journal of Primatology* **11**:173–92.

Danielsen L, Aalund O, Mazza PH, and Katz E (1989) Injuries due to deliberate violence in areas of Argentina. II. Lesions. *Forensic Science International* **42**:165–175.

Elliot-Smith G and Wood Jones F (1910) *The Archaeological Survey of Nubia Report for 1907–1908*, Volume II, *Report on the Human Remains*. Cairo: National Printing Department.

Grauer AL (1995) Analysis of the human skeletal remains from the Tremaine Complex. In J O'Gorman (ed.), *The Tremaine Site Complex: Oneota Occupation in the LaCrosse Locality, Wisconsin*. Archaeology Research Series 3. Museum Archaeology Program. State Historical Society, Wisconsin.

Grauer AL and Roberts C (1996) Paleoepidemiology, healing, and possible treatment of trauma in a Medieval cemetery population of St. Helen-on-the-Walls, York, England. *American Journal of Physical Anthropology* **100**:531–44.

Hawkes SC (1989) *Weapons and Warfare in Anglo-Saxon England.* Oxford: Oxford University Committee for Archaeology Monograph No. 21.

Hedges BE, Dimsdale JE, Hoyt DB, Berry C, and Leitz K (1995) Characteristics of repeat trauma patients, San Diego County. *American Journal of Public Health* **85**:1008–10.

Jurmain R (1989) Trauma, degenerative disease, and other pathologies among the Gombe Chimpanzees. *American Journal of Physical Anthropology* **80**:229–37.

Jurmain R (1990) Paleoepidemiology of a central California prehistoric population from CA-Ala–329: dental disease. *American Journal of Physical Anthropology* **81**:333–42.

Jurmain R (1991) Paleoepidemiology of trauma in a prehistoric central California population. In DJ Ortner and AC Aufderheide (eds.), *Human Paleopathology: Current Synthesis and Future Options*. Washington: Smithsonian Institution Press, pp. 241–8.

Jurmain R (1997) Skeletal evidence of trauma in African apes, with special reference to the Gombe chimpanzees. *Primates* **38**:1–14

Jurmain R and Bellifemine V (1997) Patterns of cranial trauma in a prehistoric population from central California. *International Journal of Osteoarchaeology* **7**:43–50.

Kilgore L, Jurmain R, and Van Gerven DP (1997) Paleoepidemiological patterns of trauma in a Medieval Nubian skeletal population. *International Journal of Osteoarchaeology* **7**:103–14.

Lambert P (1994) *War and Peace on the Western Front: A Study of Violent*

Conflict and its Correlates in Prehistoric Hunter–Gatherer Societies of Coastal Southern California. Ph.D. dissertation, University of California, Santa Barbara.

Lambert P and Walker P (1991) Physical anthropological evidence for the evolution of social complexity in coastal southern California. *Antiquity* **65**:963–73.

Leventhal AM (1993) *A Reinterpretation of Some Bay Area Shelmound Sites: A View from the Mortuary Complex at Ca-Ala-329, the Ryan Mound.* M.A. thesis, San Jose State University, San Jose, CA.

Lovejoy CO and Heiple KG (1981) Analysis of fractures in skeletal populations with an example from the Libben Site, Ottawa County, Ohio. *American Journal of Physical Anthropology* **55**:529–41.

Lovell N (1990) *Patterns of Injury and Illness in Great Apes.* Washington: Smithsonian Institution Press.

Manson JH and Wrangham R (1991) Intergroup aggression in chimpanzees and humans. *Current Anthropology* **32**:369–90.

Matthew PK, Kapua F, Soaki PJ, and Watters DAK (1996) Trauma admissions in the southern highlands of Papua New Guinea. *Australian and New Zealand Journal of Surgery* **66**:659–63.

McNamara EM (1996) Evidence of inter-personal conflict in a prehistoric Native American population from LaCrosse County, Wisconsin (Abstract). *American Journal of Physical Anthropology* **22**:164–5.

Merbs CF (1983) *Patterns of Activity-Induced Pathology in a Canadian Inuit Population.* Archaeological Survey of Canada, Paper No. 119. Ottawa: National Museums of Canada.

Milner GR, Anderson E, and Smith VG (1991) Warfare in late prehistoric west-central Illinois. *American Antiquity* **56**:581–603.

Newman RW (1957) *A Comparative Analysis of Prehistoric Skeletal Remains from the Lower Sacramento Valley.* California Archaeological Survey Reports, No. 39. Berkeley, CA: University of California.

Owsley DW, Berryman HE, and Bass WM (1977) Demographic and osteological evidence for warfare at the Larson site, South Dakota. *Plains Anthropology Memoir* **13**:119–31.

Owsley DW, Gill GW, and Owsley SD (1994) Biological effects of European contact on Easter Island. In CS Larsen and GR Milner (eds.), *In the Wake of Contact: Biological Responses to Conquest.* New York: Wiley-Liss, pp. 161–77.

Pfeiffer S (1985) Paleopathology of Archaic peoples of the Great Lakes. *Canadian Review of Physical Anthropology* **4**:1–7.

Shepherd JP, Shapland M, Pearce NX, and Scully C (1990) Pattern, severity and aetiology of injuries in victims of assault. *Journal of the Royal Society of Medicine* **83**:75–8.

Smith MO (1996) Parry fractures and female-directed interpersonal violence: implications for the late Archaic period of west Tennessee. *International Journal of Osteoarchaeology* **6**:84–91.

Strom C, Johanson G, and Nordenram Å (1992) Facial injuries due to criminal violence: a retrospective study of hospital attenders. *Medicine, Science and the Law* **32**:345–53.

Tenney J (1986) Trauma among early California populations. *American Journal of Physical Anthropology* **69**:271.

Tyson R (1977) Historical accounts as aids to physical anthropology: examples of head injury in Baja California. *Pacific Coast Archaeological Society Quarterly* **13**:52–8.

Van Gerven DP, Sandford MK, and Hummert JR (1981) Mortality and culture change in Nubia's Batn-el-Hajar. *Journal of Human Evolution* **10**:395–408.

Van Gerven DP, Sheridan SG, and Adams WY (1995) The health and nutrition of a Medieval Nubian population. *American Anthropologist* **97**:468–80.

Waldron T (1994) The human remains. In V Evison (ed.), *An Anglo-Saxon Cemetery from Great Chesterford, Essex.* York: Council for British Archaeology, Research Report 91, pp. 52–66.

Walker P (1989) Cranial injuries as evidence of violence in prehistoric southern California. *American Journal of Physical Anthropology* **80**:313–23.

Webb S (1995) *Paleopathology of Aboriginal Australians.* Cambridge: Cambridge University Press.

Wells C (1964) *Bones, Bodies, and Disease.* New York: Preager.

Wilkinson RG and Van Wagenen KM (1993) Violence against women: prehistoric skeletal evidence from Michigan. *Mid-continent Journal of Archaeology* **18**:190–216.

Willey P (1990) *Prehistoric Warfare on the Great Plains: Skeletal Analysis of the Crow Creek Massacre Victims.* New York: Garland Press.

3

Osteoporosis in the bioarchaeology of women

DAVID S. WEAVER

Osteoporosis is a topic of considerable interest today, with a tremendous amount of media coverage and varied marketing strategies geared toward the public. The topic has led to increased interest in the history and patterns of the condition. There is a substantial number of bioarchaeological studies of osteoporosis. The enthusiasm concerning the condition has given emphasis to a series of important questions. For example, have women suffered from osteoporosis throughout human history? What were the causes of osteoporosis in the past? Can we learn about modern populations from examining archaeological samples? Is the modern condition relevant to an understanding of osteoporosis in the past? For researchers, serious questions remain concerning the immediate and potentially evolutionary causes of this condition. However, concern is also warranted when modern patterns of osteoporosis are used to explore the past. In this chapter some of the pitfalls and promises of studying osteoporosis in clinical and archaeological populations will be discussed, and the assumptions and conclusions often made by skeletal biologists, paleopathologists, and bioarchaeological researchers will be examined.

To begin, it is essential to clarify some aspects of the study of human bone loss. For example, it is imperative to employ specific terminology, especially when terms imply or lead to a particular diagnosis with associated interpretations and conclusions. Early authors observing lower than expected amounts of bone, or low quality of bone, in archaeological samples often used the term 'osteoporosis'. Although using a single term does have advantages, complex situations usually are not reducible to single terms without forfeiting information and inferential strength. This is especially true when a term is adopted from another field of study. In these instances, the use of the term often does not keep pace with the findings and developments in the original field. For instance, the term 'osteoporosis' is currently used more specifically than its original meaning of the simple observation of thin or low density bone in patient radiographs (see Albright *et al.* 1940). As diagnostic techniques, our understanding of the etiologies and physiology, and our understanding of the consequences and treatments of low bone density have become more sophisticated, researchers have recognized that there are many ways to create thin bone or low bone density (Resnick and Niwayama 1988; Coe and Favus 1992), and therefore more specific terminology has been developed.

Because the terms osteoporosis and low bone density (also known as osteopenia) are often confused, it may prove useful to distinguish between the more generic observation of osteopenia and the specific modern clinical syndrome of osteoporosis (see White and Armelagos 1997 for a more extensive discussion). Osteopenia is simply defined as the presence of less than normal amounts of bone (Bilezikian *et al.* 1996). Although osteopenia is a necessary condition for osteoporosis, alone it may or may not have important consequences for an individual. The condition may be transient and is potentially reversible, which occurs during states of high bone turnover. Osteopenia has many forms and etiologies. Thus, the term is usually used in conjunction with an etiology, as in calcium deficiency osteopenia (Anderson and Garner 1995).

While osteopenia may be indicative of an individual's bone metabolism and perhaps their state of health, osteoporosis is a clinical syndrome whereby bone fracture results from less than normal amounts or quality of bone (Alvioli 1993; Mundy 1995). It is a complex condition, classified as primary, secondary, age-related, or postmenopausal. These classifications, however, are not mutually exclusive. Osteoporosis is usually irreversible (Mundy 1995). Since both osteopenia and osteoporosis may have a wide range of consequences for an individual, from no appreciable effect to substantial morbidity and death (Repa-Eschen 1993), each is important to observe and evaluate in bioarchaeological studies.

Observing the amount and architecture of bone presents an appealing opportunity for paleopathologists and bioarchaeologists to interpret the past. Inferences about physiology, nutrition, lifestyle, sex, and associated behavior are often made in these studies. The inferences and arguments range from general statements (as when discussing general 'stressors' *sensu* Goodman *et al.* 1988) to very specific diagnoses and conclusions. As is true for many bioarchaeological conditions, such as porotic hyperostosis (see Stuart-Macadam 1992 and Chapter 4), interpreting the presence and patterns of lesions is complex. Nonetheless, a number of assumptions are commonly used among bioarchaeologists as they explore the presence of bone loss in past populations. Understanding the bases and problems of these assumptions is critical to the discussion of the pitfalls and promise of exploring osteopenia and osteoporosis in human skeletal populations.

Bioarchaeological expectations

Bioarchaeological expectations about osteopenia and osteoporosis can be summarized as follows:

(1) Bone amounts and quality can be measured reliably in archaeological populations.
(2) Measured amounts of bone or estimates of bone quality will be bioarchaeologically meaningful.
(3) Diet, physiology, and other 'stressors' can be detected by the presence of osteopenia and osteoporosis.
(4) Osteopenia and osteoporosis will be more severe in females due to menopause, pregnancy, and lactation.

From what we know about modern osteopenia and osteoporosis, we can evaluate each of the bioarchaeological expectations. We also can make predictions about the accuracy of the expectations and about the results we would need to see in bioarchaeological studies to test each of the expectations.

Expectation 1: bone amounts and quality can be measured reliably in archaeological populations

There is no question that we can measure the amount of bone in archaeological specimens. However, methods vary in their accuracy and relevance. Simply weighing bones is the least accurate and the least informative way to evaluate amount and quality of bone. Postmortem changes, including soil and water infusion and subsequent mineralization, will be variable and have substantial influence on bone weight. Therefore, weighing bone will shed little light on bone quantity, density, or quality. Radiography (X-rays) or dual photon absorptiometry of whole bones are also commonly used (see Rothschild 1992), from which percentage of cortical area values are estimated. Although these techniques are inexpensive and quick, there are a number of problems with these methods. Bone curvature, variations in bone density and radio-opacity, as well as postmortem changes, all severely limit the utility of whole bone radiographic methods. Even including density wedges in radiographs is not likely to improve our knowledge unless we understand and control for post-mortem change. Similar to the increase in bone weight, thin or 'light' bone can result from postmortem modifications, due to the effects of water and soil infusion, plant and fungal infestation, and recovery and preparation techniques (Schultz 1997). Given these problems, whole bone radiography should be viewed as a screening technique, by which we identify bone for further study, and not as a specific diagnostic tool. We can improve on whole bone techniques by using a couple of methods that are becoming more common in bioarchaeology – bone microradiography and bone histology.

Microradiography entails more expense and time than whole bone radiography, as prepared bone sections and relatively expensive preparation and testing equipment are required. Inclusion of an appropriate density standard is also essential and microradiographic exposures need to be carefully monitored. This method, however, allows direct measurement of the cross-sectional area (and other characteristics) of bone and reveals density differences. Techniques such as bone histology and histomorphometry also provide a great deal of information about the amount and character of bone. Unfortunately, the substantial expense in preparation and equipment, as well as the need for well-trained observers, has limited the use of these techniques. These are, however, the only methods that can provide information about bone turnover and bone metabolism in bioarchaeology.

Expectation 2: measured amounts of bone or estimates of bone quality will be bioarchaeologically meaningful

What are some of the problems with analogies between modern osteopenia and osteoporosis and recognizably thin or poor bone quality and quantity in archaeological populations? First, modern bone conditions are truly multifactorial, both in their etiology and their manifestations (Resnick and Niwayama 1988; Coe and Favus 1992; Favus 1996). It is unlikely, therefore, that any single etiology will be adequate to explain osteopenia, much less osteoporosis, in a bioarchaeological setting. Hidden multiple etiologies and manifestations surely will complicate bioarchaeological interpretations.

Second, as mentioned above, the frequently used traditional radiographic and direct measures of bone may not be adequate to assess osteopenia or osteoporosis across populations and time. Modern standards for the conditions are population-specific and rely on the age, sex, habitus, and other variables of the sample (Alvioli 1993). Differences in radiographic techniques and geometric influences on bone measurements will produce only relative values – values that will prove very difficult to compare across samples even when comparisons within samples can be evaluated. Similarly, direct measurement of bone cross-sections can yield estimates of the amount of bone present, but will not yield information about the quality or density of that bone. As Chestnut (1993) has stated, simply recognizing 'thin bone' and even 'poor bone quality' within a population are seldom reliable predictors of fracture risk or bone changes associated with physiological conditions. Bone geometry, especially cross-sectional geometry at important skeletal loading sites, is as important as the amount or quality of bone in determining the skeletal health of the individual (Frost 1985; Ruff 1992).

Lastly, it must be recognized that clinically significant osteopenia and osteoporosis may be a largely modern phenomenon (Eaton and Konner 1985; Neese and Williams 1994; but also see Dequeker *et al.* 1997; Drezner 1997). It is interesting that despite notable exceptions (e.g., White and Armelagos 1997), modern patterns of thin bone and the characteristic wrist, vertebral, and proximal femur fracture patterns of osteoporosis, are not widespread in bioarchaeological samples. Perhaps, theoretically, widespread clinically significant osteopenia or osteoporosis over long periods of time is evolutionarily counterintuitive (Neese and Williams 1994). We should expect that a person's skeleton should be well adapted to personal demands and that, in general, only extraordinary circumstances should yield skeletal failure (Frost 1986). Indeed, in most cases it might be necessary to view bioarchaeological instances of thin bone or poor bone quality as special cases requiring special explanation, rather than as a general measure of human adaptation.

If we could accurately measure the amount (and even the quality) of bone in bioarchaeological studies, does this necessarily mean that we could identify and interpret the results in a bioarchaeologically meaningful way? The problem with our lack of adequate standards, and the consequent need to concentrate on extreme cases and results, may be tempting us prematurely to make specific diagnoses and interpretations in archaeological specimens. Unfortunately, the relative simplicity of the bone remodeling system works against us. There are only two bone cell types doing the actual work of bone remodeling – osteoblasts and osteoclasts. The role of osteocytes (Hall 1991) remains largely unknown. There are thus only two possible cell behaviors – resorption by osteoclasts (Hall 1991) and formation by osteoblasts (Hall 1990). Hence, there is very little diagnostic precision inherent in bone unless characteristic bone lesions or patterns occur. In osteopenia and osteoporosis, bone loss is due to a skewed bone maintenance system that favors resorption over formation. Its presence provides almost no specific diagnostic information. In fact, osteopenia and osteoporosis can have many specific etiologies, with many different bioarchaeological explanations and interpretations. Complementary information, such as hormone levels, vitamin D status, or specific risk factors within a population, is essential in the determination of etiology. Ecological and behavioral information is also critical, but found in only a few bioarchaeological contexts (such as in White and Armelagos 1997). Usually, we can only say that the bones under study display less bone, or lower bone quality than we expected, and that the findings result from higher (or lower) than 'normal' bone turnover. Of course, even this general level of interpretation begs the question of what might be 'normal'. Hence, a great deal of ancillary work still needs to be undertaken, including basic osteological and paleodemographic analyses, biomechanical studies, extensive archaeological research, and research strategies using large samples with careful and appropriate methodology. With these criteria being met, we may be able to move beyond the general and essentially trivial observation of low amounts or poor bone quality to try to explain the observation in a bioarchaeologically useful way.

Expectation 3: diet, physiology, and other stressors can be detected by the presence of osteopenia and osteoporosis

Many studies have attributed observed bone loss, or less than expected amounts or quality of bone, to diet and other stressors. If we are going to use modern bone to understand archaeological specimens and derive inferences and offer explanations, we should be clear about the modern syndrome. If we assume that the relevant bone cells and bone related physiology have remained similar throughout the history of *Homo sapiens,* then we can concentrate on modern risk factors and causes for modern osteopenia and osteoporosis as a means towards understanding bioarchaeological findings. In modern populations the most commonly cited risk factors for osteopenia and osteoporosis are growth and development, nutrition, habitus, endocrine and other hormone status, reproductive status, aging, disease, and biological variability (Alvioli 1993; Mundy 1995).

Bone growth, development, and remodeling, along with the acquisition of peak bone mass, are essential but poorly understood factors in the risk of adult osteopenia or osteoporosis (see Stini 1990). Malnutrition, calcium deficiency, and the diseases associated with malnutrition, all have detrimental effects on achieving full genetically-established peak bone mass. The peak bone mass achieved (usually early in adulthood), sets the point from which subsequent net bone loss occurs. Bone loss occurs as a function of aging and as a result of other risk factors, and contributes to an individual's susceptibility to osteopenia or osteoporosis. If female children are nutritionally disadvantaged compared to male children, affecting the establishment of peak bone mass, then females would be more liable to develop osteopenia or osteoporosis earlier in life. Determining this pattern in archaeological populations is difficult, but is possible if the paleo-environment and paleodiet, along with social conditions of the population can be reconstructed. However, it must be stressed that females, often by virtue of being smaller in size, might accumulate lower peak bone mass, and therefore be more liable to osteopenia and osteoporosis than males (Alvioli 1993).

Most bioarchaeological interpretations of osteopenia and osteoporosis have focused on the presence of calcium in skeletal tissue because of its major contribution to the mineral structure of bone, the substantial understanding of its roles and interactions in the human body, and the fact that it remains in bone after death and decomposition. Unfortunately, except in defective mineralization syndromes, calcium is constant per unit of hydroxyapatite in bone (Anderson and Garner 1995). Thus, proportional studies of the amount of calcium in bone samples do not promise much information about osteopenia or osteoporosis. This is especially true if diagenesis has occurred, affecting the amount of surviving bone tissue (Schultz 1997). However, if diagenetic loss does not occur, the absolute amount of calcium in a bone sample should be at least a rough indicator of bone density and amount.

The relationship between calcium and diet is also frequently explored in bioarchaeological studies. This is likely due to the known modern links between nutrition and the presence of osteopenia and osteoporosis. Low calcium intake, especially during growth and other periods of high bone turnover, may yield osteopenia, which in time may progress to osteoporosis. Effective calcium intake is greatly influenced by dietary calcium availability and absorption (Stini 1990; Heaney 1993; Anderson and Garner 1995). The availability of calcium is a function of the calcium content in foods, which can vary widely (Anderson and Garner 1995). Calcium absorption is affected by the proportion of fat and protein in the diet, the level and activity of vitamin D, the presence of concurrent disease, and the general health of the individual. The presence of calcium binding or chelating compounds, including phytates (Stini 1990; Anderson and Garner 1995 and many others) is also important to calcium absorption. The complexity of the body's interactions and mechanisms to insure calcium balance suggests that maintaining adequate calcium balance is critical (Anderson and Garner 1995). In evolutionary terms, serious physiologically induced calcium imbalances should be quite rare, given that the acute cost of unsatisfactory calcium homeostasis is death (Guyton 1976). It seems unlikely, therefore, that human populations withstood extensive periods when diets were severely deficient in calcium. Rather, populations experiencing more marginal or periodic deficiencies are more likely to survive the episode(s) and possibly to display the skeletal effects.

The effects of diet on calcium absorption might have been important to the health status in some archaeological populations. The effects might also have impacted women differently than men. For instance, low fat diets inhibit calcium absorption. A number of authors (e.g., Larsen 1984) have argued that women might have had less satisfactory diets than men, particularly under some agricultural regimes. Women might obtain less vitamin D, or might convert vitamin D less effectively, due to dietary or activity-based differences between women and men, thus inhibiting calcium absorption. High protein diets are also implicated in reduced calcium absorption (see Stini 1990). Although seemingly promising, clear cases of differential protein intake between women and men in archaeological contexts have not been successfully recognized independent of the presence of differing amounts of bone quality or quantity in the skeletal samples. Calcium binding compounds, including the phytates that are found in many grain crops (Anderson and Garner 1995; Stini 1990), also inhibit calcium absorption. If women have larger grain components in their diets in some agricultural settings (see Larsen 1984), then presumably they should be at a disadvantage for calcium absorption.

Parasites (particularly intestinal parasites), also affect calcium absorption and provoke calcium loss directly and indirectly. They also impact iron and other nutrients (Hoeprich 1977). Calcium absorption is decreased, and calcium loss is increased, with the presence of diarrhea and other intestinal illnesses brought on by the presence of parasites. Importantly, disease (whether parasitically induced or not) and malnutrition are often synergistic in their adverse effects on the body and on bone (McElroy and Townsend 1979; Harrison *et al.* 1988). If women are limited to a 'home area', or to areas of higher parasite exposure, they might have heavier parasitic loads. This can reduce calcium uptake or increase the intestinal loss of calcium. In sedentary environments involving full domestication, where subsistence roles for women and men are similar, it seems likely that parasitic effects on calcium absorption should be similar (Reinhard 1988). However, if women experience differences in exposure or susceptibility to intestinal parasites and illness, then they might have higher calcium deficits together with higher disease loads. This might impact their skeletal system.

Chronic diseases, probably because of their effects on general health, can also provoke osteopenia and even osteoporosis (Resnick and Niwayama 1988). Many metabolic diseases have adverse affects on bone (Resnick and Niwayama 1988). Inflammatory diseases, such as lupus and some arthroses (rheumatoid arthritis, for example; see Rothschild *et al.* 1997) can result in osteopenia and even osteoporosis (Resnick and Niwayama 1988). Metastatic disease, particularly cancers that increase osteoclast activity, can create both local and general osteopenia and can result in pathological fractures that may mimic typical osteoporotic fractures (Resnick and Niwayama 1988; Mundy 1995). The lesson here is that as long as various diseases can adversely affect bone, we need to be careful to diagnose the disease independently of the presence of osteopenia or osteoporosis. This will insure that any bone loss (a general result) is not diagnosed as a specific condition (the disease in question), and that specific disease conditions are not confused with general bone loss. In bioarchaeological populations, this need is often difficult to meet.

In the evaluation and interpretation of bone loss or low bone quality, the role of human biological variability is also increasingly recognized as a factor that needs to be considered (Villa 1994). For example, there are substantial, yet unexplained differences in vitamin D metabolism and parathryoid hormone form and function between biological groups (Cosman *et al.* 1997; Riggs 1997). It has been suggested that these differences explain the observation of higher bone density in black women than in other biological groups (Parisien *et al.* 1997), even when age, habitus, and other risk factors are statistically controlled. It is probable that different human groups will display different results in archaeological populations as well. However, standards for evaluating osteopenia must be sample or population specific if meaningful interpretations are to be created.

Habitus – a suite of personal, cultural, and biological behaviors – can affect peak bone mass and the subsequent risk of osteopenia or osteoporosis. Physical work is one of the most important aspects of habitus. The skeleton clearly adapts to work demands, both in form and amount of bone, particularly during growth and development (Frost 1996). The skeleton also adapts to work demands, usually within narrower limits (set by growth and peak bone mass), later in life. Thus, exercise can at least partially prevent or delay bone loss in older persons (Stini 1990; Prince *et al.* 1995; Ward *et al.* 1995). Known adverse components of habitus include smoking, alcohol, and drug use. Still other less common environmental toxins (such as lead), and behaviors such as stringent dietary restrictions (particularly restricting fat energy and calcium), can affect skeletal tissue. As for many of the risk factors, the key to understanding different patterns of bone quantity or quality lies in being able to identify different patterns of behavior or different patterns of exposure to toxins between females and males. Such identification depends on archaeological and other sources of information.

In conclusion, even if we accept that observations of bone are accurate and appropriate, there remain serious problems with most of the specific interpretations of the findings. This is primarily due to the overwhelming complexity of the 'causes' of osteopenia and osteoporosis. For instance, because osteopenia and osteoporosis almost always are multifactorial, because our samples are cross-sectional by nature, because our bioarchaeological samples are regionally or locally specific, and because complementary and independent bioarchaeological information may not be available, it is very unlikely that specific etiologies or even population patterns will be fully understood. Interpretation of low bone amounts or poor bone quality in the archaeological record, particularly as it pertains to women, is probably best stated as resulting from general 'stressors' rather than being attributed to specific causes. Of course, when more specific interpretations can be offered and supported, those interpretations will provide important information for bioarchaeologists.

Expectation 4: osteopenia and osteoporosis will be more severe in females due to pregnancy, lactation and menopause

One circumstance in which we clearly see biological differences between females and males is in the expression and effects of endocrine hormones. The primary hormone affecting bone in females is estrogen, although progesterone and other steroid hormones also play a role. Insufficient estrogen levels, at any time in a woman's life, will increase bone remodeling rates and continually result in loss of bone (Turner *et al.* 1994). Both hormone expression and deprivation have their effects on the skeleton, and if we use patterns of modern osteopenia and osteoporosis to interpret the past, females in archaeological populations should display an earlier and greater risk of bone loss than males. The two most common ways estrogen levels are reduced in women, and the two ways most relevant to bioarchaeology, are nutritional effects and postmenopausal reduction of estrogen production.

Amenorrhea, the absence or abnormal cessation of the menstrual cycle, is well known among young women who engage in heavy aerobic exercise and among women undertaking radical dieting regimes (Seeman *et al.* 1992). While the exact mechanisms provoking the onset of dietary or exercise-induced amenorrhea are not known (Frisch 1978; Drinkwater 1987), there appear to be recognizable alterations to bone (Herzog *et al.* 1993). It is often assumed that nutritionally-induced amenorrhea could contribute to patterns of low bone quality and quantity in some skeletal populations where the diets of women might be deficient of energy, fat, calcium, or any combination of these dietary components. However, modern examples of these types of amenorrhea are culture-specific. They are often induced by excessive exercise regimens, anorexia, and binge dieting. Very little is known about marginal rather than radically deficient diets. It is not clear, therefore, whether modern examples of nutritional amenorrhea can serve as analogies towards explaining osteopenia or osteoporosis in bioarchaeological contexts.

Postmenopausal estrogen deficiency is also a well documented condition in modern women (Alvioli 1993; Mundy 1995). Bone cell responses to this condition in modern women are dramatic and highly significant for a substantial subset of women (Alvioli 1993). Superficially, it seems reasonable to postulate dramatic and significant bone responses in postmenopausal women in bioarchaeological contexts. However, several points should be considered before an uncritical acceptance of, and perhaps undue enthusiasm for this conclusion is adopted.

First, as suggested by Wood *et al.* (1992), it is difficult for researchers to ascertain many of the dynamics of an archaeological population's demography. It is even more difficult, if not impossible, to determine how many women in a given sample actually entered the postmenopausal period. This is particularly true if we take pains to avoid the circular argument that low bone mass implies a female, that females with low bone mass are postmenopausal, and so a skeleton with low bone mass probably represents a postmenopausal female. The fact is that without using independent age estimations and adequate paleodemographic techniques, we often cannot ascertain whether postmenopausal bone loss was an important factor in women's lives. As a further complication, age-related osteopenia and even osteoporosis can affect older males (Jackson 1996). Hence, we cannot infer that a skeleton of an older person is female simply on the basis of low bone mass.

Second, we cannot assume that menopause had dramatic effects on prehistoric women's physiology. It is likely that the number of menstrual cycles experienced by many modern women is, in an evolutionary sense, highly unusual (see Stuart-Macadam, Chapter 4 for discussion of effects of increased menstrual cycles on iron deficiency anemia). Women in urban societies today can experience as many as 450 menstrual cycles (Eaton *et al.* 1994). Estimates for modern foraging societies cluster around 50 cycles in a woman's lifetime (Short 1976; Sperling and Beyene 1997), with about 100 cycles estimated for women among the Dogon, a non-western agricultural society (Strassmann 1997). Like many other hormone receptors, the estrogen receptors on bone cells are probably highly sensitized by repeated exposure to estrogen (Turner *et al.* 1994). Hence, it is likely that in modern women the dramatic down regulation of osteoblasts, in response to the postmenopausal withdrawal of estrogen, is at least partly due to the unusually high number of menstrual cycles experienced by these women.

Parity also plays a role in the maintenance of bone tissue after menopause (Galloway 1988; Stini 1990). Nulliparity among modern women is positively associated with postmenopausal osteoporosis. This may be because nulliparous women have more menstrual cycles than multiparous women, thus exposing their cell estrogen receptors to more frequent estrogen doses. Of course, nulliparous women could also have been estrogen deficient for all or part of their adult lives, predisposing them to osteopenia, even before menopause, and to osteoporosis later in life.

To further complicate matters, it is possible that plant phytoestrogens (Setchell *et al.* 1984), such as those found in soy and maize (Arjmandi *et al.* 1996), may have prophylactic effects in postmenopausal women. Current research seeks to understand the actions of phytoestrogens in postmenopausal women. It is possible that phytoestrogen intake may partly forestall postmenopausal bone loss and cardiovascular changes. If this is true, we should not expect dramatic postmenopausal bone loss among women in bioarchaeological contexts where phytoestrogens were important dietary components.

A number of bioarchaeological studies have attributed low bone mass or thin cortical bone to the effects of pregnancy and/or lactation (Dewey *et al.* 1969; see White and Armelagos 1997 for an informative discussion). Modern studies have failed to support this argument, except when pregnancy or lactation are associated with dietary deficiencies, or when lactation is continued for several years (Anderson and Garner 1995; Sowers 1996). Indeed, in women with satisfactory nutrition and health, pregnancy and lactation are associated with improved bone mass and bone quality (Sowers 1996). This outcome may be due to the increased bone remodeling (often with transient and reversible osteopenia) that occurs during pregnancy and lactation. In these instances, more bone is resorbed and subsequently rebuilt in pregnant, lactating women than in non-pregnant, non-lactating women. In effect, pregnancy and lactation under adequate nutrition 'reset' much of the skeleton with newer, better quality bone (Sowers 1996).

Thin bone and low bone mass might be recognizable in bioarchaeological samples, especially in situations where maternal nutrition is suspect and/or when long lactation periods can be inferred. Modern correlates to this association can be found in Inuit women, where osteopenia and even osteoporosis has been attributed to long lactation periods and excess dietary protein (Mazess and Mather 1975). Conversely, Lazenby (1997) has offered the provocative alternative that various metabolic aspects of cold adaptation might contribute to bone loss in these women. White and Armelagos (1997) summarize a number of other ways in which pregnancy and lactation could have adverse effects on bone. Therefore, there are a number of bioarchaeological circumstances under which pregnancy and lactation might be detrimental to the quantity and quality of bone. It is interesting to note, however, that typical osteoporotic fractures are not often seen in bioarchaeological samples, even in women with thin bone or low bone mass. This may indicate that women with low bone mass in these archaeological samples found that pregnancy and lactation provided beneficial effects on the quality of their bone. It may also be true that their life experiences helped to create a skeleton with appropriate characteristics and bone geometry (Frost 1986), reducing their liability to fracture even in the face of reduced amounts or poor quality of bone. The key point is that we should not automatically associate bone loss with pregnancy and lactation, but must independently establish dietary and behavioral reasons to expect that pregnancy and lactation, within a particular group of women, would have had adverse effects on bone.

Modern studies have shown that pregnancy and lactation should not result in permanent bone deficits unless poor nutrition or other complicating factors prevent skeletal remodeling and recovery from the bone loss inherent in pregnancy and lactation. This situation makes evolutionary sense, in that pregnancy and lactation should not, in the interest of long-term maternal fitness, place a woman at an appreciable disadvantage (Neese and Williams 1994). It is difficult to argue that maternal nutrition was unsatisfactory throughout human evolution, unless we assert that osteopenia and osteoporosis were of little consequence to females. Indeed, an appropriate evolutionary expectation would be that calcium and bone-related nutritional requirements were within acceptable limits, making skeletal pathology, with its associated survival risks, less likely to occur over most of human evolution.

Our current widespread dependence on agriculture is a relatively recent event (see Cohen and Armelagos 1984). So too, is our longevity and the many modern risk factors for osteopenia and osteoporosis (Alvioli 1993; Drezner 1997). Since most bioarchaeological populations come from this recent period of human evolution, studies exploring the effects of pregnancy and lactation, and other risk factors, are crucial to our understanding of how this subsistence strategy influences bone quality and quantity. The studies are not, however, representative of human adaptive processes over the course of human evolution. Unless complementary and supporting bioarchaeological evidence is available, we should not expect pregnancy and lactation, or any of the many modern factors influencing bone growth, development, maintenance, and demise, to have routinely caused low bone amounts or poor bone quality among prehistoric women.

Similarly, there is no *a priori* reason to expect appreciable differences in bone quantity and quality between the sexes. Our expectations concerning bone have been prejudiced by modern cases of postmenopausal and age-related osteopenia and osteoporosis. We simply do not know the impact that postmenopausal changes had on the skeletal systems of prehistoric women. This is not to deny that females faced different and at times more serious risk factors than males in the past. These might have included different diets, the effects of pregnancy and lactation under marginal nutrition, and risks inherent to different socio-cultural behaviors and lifestyles. However, until we can adequately ascertain the age of older females and males from skeletal material, and can identify different risk factors for females and males through independent bioarchaeological information, it is clearly premature to attribute female–male differences in bone amounts or quality to postmenopausal causes.

Conclusion

If we are going to gain a better understanding and interpretation of low bone quantity or poor bone quality in the bioarchaeology of women, what remains to be done? Although the absence of cells and their associated biochemistry in most archaeological samples will continue to limit our interpretations, we should continue to improve our methods for identifying and describing bone tissue, and for recognizing events and processes in bone. Schultz (1997) has contributed greatly to this end, but more refinement, both in methods and interpretation, is still needed. Without an understanding of individual or population consequences of osteopenia or osteoporosis, there is little to be gained by merely describing low amounts or poor quality of bone in archaeological samples. This is because the mere fact that the presence of low bone quantity or quality exists does not provide useful information about the nature of bone fractures, bone mechanical performance, bone-related metabolism, or disease. Only methods that examine bone cell behavior and the consequences of that behavior, such as histology, histomorphometry, and perhaps microradiography, actually hold promise for diagnoses and explanations of low bone amounts or poor bone quality.

We also need to refine our understanding of the consequences of low bone quantity and quality, and not assume specific outcomes. We should carefully examine our tendency to create simplistic and unicausal 'explanations' for multifactorial syndromes such as osteopenia and osteoporosis. As always, we should ensure that our interpretations and explanations stand on their own and are not dependent on circular arguments.

In general, using the analogy of modern human osteopenia and osteoporosis in bioarchaeology has yielded rewards for researchers working to understand the bioarchaeology of women. We need to continually examine, update, and improve our models of osteopenia and osteoporosis based upon modern medical syndromes. There are good reasons to think that the analogy will continue to be useful. This, along with the synthesis of information from archaeology, paleonutrition, paleodemography, ethnohistory, and other fields will continue to strengthen our ability to develop hypotheses and adequately test them.

However, we should be mindful that exploring multiple lines of evidence is only useful if they remain independent of each other. Indeed, similar constraints exist for almost every bioarchaeological problem.

References

Albright F, Bloomburg F, and Smith PH (1940) Postmenopausal osteoporosis. *Transactions of the Association of American Physicians* **55**:298–305.

Alvioli LV (Ed.) (1993) *The Osteoporotic Syndrome: Detection, Prevention, and Treatment.* New York: Wiley-Liss.

Anderson JJB and Garner SC (Eds.) (1995) *Calcium and Phosphorus in Health and Disease.* Boca Raton: CRC Press.

Arjmandi BH, Alekel L, Hollis BW, Amin D, Stacewicz-Sapuntzakis M, Guo P, and Kukreja SC (1996) Dietary soybean protein prevents bone loss in an ovariectomized rat model of osteoporosis. *Journal of Nutrition* **126**:161–7.

Bilezikian JP, Raisz LG, and Rodan GA (Eds.) (1996) *Principles of Bone Biology.* San Diego: Academic Press.

Chestnut CH (1993) Noninvasive methods for bone mass measurement. In LV Alvioli (ed.), *The Osteoporotic Syndrome: Detection, Prevention, and Treatment.* New York: Wiley-Liss, pp. 77–87.

Coe FL and Favus MJ (Eds.) (1992) *Disorders of Bone and Mineral Metabolism.* New York: Raven Press.

Cohen MN and Armelagos GJ (Eds.) (1984) *Paleopathology at the Origins of Agriculture.* Orlando: Academic Press.

Cosman F, Morgan DC , Nieves JW, Shen V, Luckey MM, Dempster DW, Lindsay R, and Parisien M (1997) Resistance to bone resorbing effects of PTH in black women. *Journal of Bone and Mineral Research* **12**:958–66.

Dequeker J, Ortner DJ, Stix AI, Cheng X-G, Brys P, and Boonen S (1997) Hip fracture and osteoporosis in a XIIth dynasty female skeleton from Lisht, Upper Egypt. *Journal of Bone and Mineral Research* **12**:881–8.

Dewey JG, Armelagos GJ, and Bartley M (1969) Femoral cortical involution in three Nubian archaeological populations. *Human Biology* **41**:13–28.

Drezner MK (1997) Osteoporosis, a disease of the age(d)s. *Journal of Bone and Mineral Research* **12**:880.

Drinkwater BL (1987) Exercise-associated amenorrhea and bone mass. In AF Roche (ed.), *Osteoporosis: Current Concepts.* Columbus: Ross Laboratories, pp. 42–6.

Eaton SB and Konner ML (1985) Paleolithic nutrition. *New England Journal of Medicine* **312**:283–9.

Eaton SB, Pike MC, Short RV, Lee NC, Trussell J, Hatcher RA, Wood JW, Worthman CM, Blurton Jones NG, Konner MJ, Hill KR, Bailey R, and Hurtado AM (1994) Women's reproductive cancers in evolutionary context. *Quarterly Review of Biology* **69**:353–67.

Favus MJ (Ed.) (1996) *Primer on the Metabolic Bone Diseases and Disorders of Mineral Metabolism*, 3rd edn. Philadelphia: Lippincott-Raven.

Frisch R (1978) Population, food intake and fertility. *Science* **199**:22–30.

Frost HM (1985) The 'new bone': some anthropological potentials. *Yearbook of Physical Anthropology* **28**:211–26.

Frost HM (1986) *Intermediary Organization of the Skeleton.* Boca Raton: CRC Press.

Frost HM (1996) Bone development during childhood. A tutorial (some insights of a new paradigm). In E Schonau (ed.), *Paediatric Osteology: New Developments in Diagnostics and Therapy.* Amsterdam: Elsevier, pp. 3–39.

Galloway A (1988) *Long Term Effects of Reproductive History on Bone Mineral Content in Women.* Doctoral Dissertation. Tucson: University of Arizona.

Goodman AH, Thomas RB, Swedlund AC, and Armelagos GJ (1988) Biocultural perspectives on stress in pre-historic, historical and contemporary population research. *Yearbook of Physical Anthropology* 31:169–202.

Guyton AC (1976) *Textbook of Medical Physiology.* Philadelphia: W.B. Saunders.

Hall BK (1990) *Bone.* Volume 1: *The Osteoblast and Osteocyte.* Caldwell: Telford Press.

Hall BK (1991) *Bone.* Volume 2: *The Osteoclast.* Caldwell: Telford Press.

Harrison GA, Tanner JM, Pilbeam DR, and Baker PT (1988) *Human Biology.* Oxford: Oxford University Press.

Heaney RP (1993) Prevention of osteoporotic fracture in women. In LV Alvioli (ed.), *The Osteoporotic Syndrome: Detection, Prevention, and Treatment.* New York: Wiley-Liss, pp. 89–107.

Herzog W, Minne H, Deter C, Leidig G, Schellberg D, Wuster C, Gronwald R, Sarembe E, Kroger F, Bergmann G, Petzold E, Hahn P, Schepank H, and Ziegler R (1993) Outcome of bone mineral density in anorexia nervosa patients 11.7 years after first admission. *Journal of Bone and Mineral Research* 8: 597–605.

Hoeprich PD (1977) *Infectious Diseases.* Hagerstown: Harper & Row.

Jackson JA (1996) Osteoporosis in men. In MJ Favus (ed.), *Primer on the Metabolic Bone Diseases and Disorders of Mineral Metabolism.* Philadelphia: Lippincott-Raven, pp. 283–8.

Larsen CS (1984) Health and disease in prehistoric Georgia: the transition to agriculture. In MN Cohen and GJ Armelagos (eds.), *Paleopathology at the Origins of Agriculture.* Orlando: Academic Press, pp. 367–92.

Lazenby RA (1997) Bone loss, traditional diet, and cold adaptation in arctic populations. *American Journal of Human Biology* 9:329–41.

Mazess RB and Mather WE (1975) Bone mineral content in Canadian Eskimos. *Human Biology* 47:45–63.

McElroy A and Townsend PK (1979) *Medical Anthropology.* North Scituate: Duxbury.

Mundy GR (1995) *Bone Remodeling and its Disorders.* London: Martin Dunitz.

Neese RM and Williams GC (1994) *Why We Get Sick.* New York: Times Books.

Parisien M, Cosman F, Morgan D, Schnitzer M, Liang X, Nieves J, Forese L, Luckey M, Meier D, Shen V, Lindsay R, and Dempster DW (1997) Histomorphometric assessment of bone mass, structure, and remodeling: a comparison between healthy black and white premenopausal women. *Journal of Bone and Mineral Research* 12: 948–57.

Prince R, Devine A, Dick I, Criddle A, Kerr D, Kent N, Price R, and Randell A (1995) The effects of calcium supplementation (milk powder or tablets) and exercise on bone density in postmenopausal women. *Journal of Bone and Mineral Research* 10:1068–75.

Reinhard KJ (1988) Cultural ecology of prehistoric parasitism on the Colorado Plateau as evidenced by coprology. *American Journal of Physical Anthropology* 77:355–66.

Repa-Eschen LM (1993) Prevention and treatment of osteoporosis in our health care delivery system. In LV Alvioli (ed.), *The Osteoporotic Syndrome:*

Detection, Prevention and Treatment. New York :Wiley-Liss, pp. 191–207.

Resnick D and Niwayama G (1988) *Diagnosis of Bone and Joint Disorders.* Philadelphia: W. B. Saunders.

Riggs BL (1997) Vitamin D-receptor genotypes and bone density. *New England Journal of Medicine* **337**:125–6.

Rothschild BM (1992) Advances in detecting disease in earlier human populations. In SR Saunders and MA Katzenberg (eds.), *Skeletal Biology of Past Peoples: Research Methods.* New York: Wiley-Liss, pp. 131–51.

Rothschild BM, Hershkovitz I, Bedford L, Latimer B, Dutour O, Rothschild C, and Jellema LM (1997) Identification of childhood arthritis in archaeological material: juvenile rheumatoid arthritis versus juvenile spondyloarthropathy. *American Journal Physical Anthropology* **102**:249–64.

Ruff C (1992) Biomechanical analyses of archaeological human skeletal samples. In SR Saunders and MA Katzenberg (eds.), *Skeletal Biology of Past Peoples. Research Methods.* New York: Wiley-Liss, pp. 37–58.

Schultz M (1997) Microscopic investigation of excavated skeletal remains: a contribution to paleopathology and forensic medicine. In WD Haglund and MH Sorg (eds.), *Forensic Taphonomy: The Postmortem Fate of Human Remains.* Boca Raton: CRC Press, pp. 201–22.

Seeman E, Szmukler GI, Formica C, Tsalamandris C, and Mestrovic R (1992) Osteoporosis in anorexia nervosa: the influence of peak bone density, bone loss, oral contraceptive use, and exercise. *Journal of Bone and Mineral Research* **7**: 1467–74.

Setchell KDR, Borriello SP, Hulme P, Kirk DN, and Axelson M (1984) Nonsteroidal estrogens of dietary origin: possible roles in hormone-dependent disease. *American Journal of Clinical Nutrition* **40**: 569–78.

Short RV (1976) The evolution of human reproduction. *Proceedings of the Royal Society of London* **B195**:3–24.

Sowers M (1996) Pregnancy and lactation as risk factors for subsequent bone loss and osteoporosis. *Journal of Bone and Mineral Research* **11**:1052–60.

Sperling S and Beyene Y (1997) A pound of biology and a pinch of culture or a pinch of biology and a pound of culture? In LD Hager (ed.), *Women in Human Evolution.* London: Routledge, pp. 137–52.

Stini WA (1990) 'Osteoporosis': etiologies, prevention and treatment. *Yearbook of Physical Anthropology* **33**:151–94.

Strassmann BI (1997) The biology of menstruation in *Homo sapiens*: total lifetime menses, fecundity, and nonsynchrony in a natural-fertility population. *Current Anthropology* **38**:123–9.

Stuart-Macadam P (1992) Anemia in past human populations. In P Stuart-Macadam and SK Kent (eds.), *Diet, Demography and Disease.* New York: Aldine De Gruyter, pp. 151–70.

Turner RT, Riggs BL, and Spelsberg TC (1994) Skeletal effects of estrogen. *Endocrine Reviews* **15**: 275–300.

Villa ML (1994) Cultural determinants of skeletal health: the need to consider both race and ethnicity in bone research. *Journal of Bone and Mineral Research* **9**: 1329–32.

Ward JA, Lord SR, Williams P, Anstey K, and Zivanovic E (1995) Physiologic, health and lifestyle factors associated with femoral neck bone density in older women. *Bone* **16**:373S–8S.

White CD and Armelagos GJ (1997) Osteopenia and stable isotope ratios in bone collagen of Nubian female mummies. *American Journal of Physical Anthropolgy* **103**:185–99.

Wood JW, Milner GR, Harpending HC, and Weiss KM (1992) The osteological paradox: problems of inferring prehistoric health from skeletal samples. *Current Anthropology* **33**:343–58.

4

Iron deficiency anemia: exploring the difference

PATRICIA STUART-MACADAM

The modern clinician views iron deficiency anemia from a limited and superficial perspective – i.e., as a condition defined by laboratory test results that requires treatment, in most cases by iron supplements. It is viewed primarily as a 'women's condition', which, according to years of misinformation generated by the medical profession and pharmaceutical companies, and propagated by the media, needs only Geritol or iron tablets to correct. In fact, iron deficiency anemia is a condition that currently affects millions of people worldwide. Wintrobe (1993) has said that it is the most common organic disorder seen in clinical medicine. Anthropological data indicates that it was also common in many past human populations (Stuart-Macadam 1992). From the perspective of biological anthropology, the pattern and demography of disease in human populations, past and present, can provide clues about the biology and culture of those populations. One important aspect of demography is the influence of sex on these variables. Are there sex-related differences in the prevalence of iron deficiency anemia today? Were there sex-related differences in the past? This chapter will explore these questions from the perspective of biological anthropology. The goal of the chapter is to synthesize paleopathological data and clinical data in order to shed light on the past and to provide a framework for studies on modern populations.

The biological anthropologist has a unique view of the world, one based on biocultural, evolutionary, and cross-cultural perspectives. The biocultural perspective acknowledges that human behavior has both biological and cultural components and that there is a reciprocal relationship between the two. The evolutionary perspective recognizes that humans exist in, and adapt to, changing environments and circumstances within a temporal context, and that our evolving biology may still reflect conditions that existed thousands, if not millions, of years ago. The cross-cultural perspective acknowledges that customs, ideas, and behaviors vary among cultures, and that understanding these variations can provide clues to interpreting biocultural data, such as health and disease patterns. How can this view contribute to an understanding of the medical condition known as iron deficiency anemia? The biological anthropologist views iron deficiency anemia not just as a medical condition to be treated, but as a condition that has existed through much of human history that can be investigated in the context of a biocultural, evolutionary, and cross-cultural framework. This approach can provide some fascinating insights into the story of iron deficiency anemia.

What is iron deficiency anemia? It is currently defined as an anemia that is associated with laboratory evidence of iron depletion provided by one or more of the following test results: low serum ferritin (iron storage protein) concentration; low transferrin (iron transport protein) saturation; or an elevation in the erythrocyte protoporphyrin level (Early and Wotecki 1993). It is important to understand that anemia is merely a *symptom* and not a disease; in clinical medicine it is considered to be an objective sign of the presence of an underlying disorder or disease (Wintrobe 1993). The causes of anemia include inadequate diet, poor iron absorption, increased iron utilization, blood loss, or infection (Florentino and Guirriec 1984). Although it is often assumed that a diet low in iron is the primary cause of iron deficiency anemia, it is rarely caused by poor diet in adults and older children who are not obviously malnourished (Lux 1995). In fact, it is a cardinal rule in diagnosis that pathological blood loss should be considered the cause of iron deficiency until proved otherwise (Bridges and Seligman 1995).

Without enough iron for hematopoiesis (the production of red blood cells), the bone marrow tries to compensate by increasing blood cell production to about twice the normal rate. However, this results in hypochromic (lacking in color) and microcytic (smaller than normal) red blood cells. The signs and symptoms of iron deficiency anemia are variable; for example, it is not uncommon for a person to be completely asymptomatic, even in severe cases of iron deficiency anemia. As well, the symptoms often show a poor correlation with the severity of anemia, at least above hemoglobin levels of 6 to 7 g/100 ml of blood, which seems to be the critical level for expression of symptoms (Wadsworth 1975). An affected individual can experience general fatigue, weakness, light-headedness, headaches, dyspnea (difficult breathing), palpitations, and paresthesias (abnormal spontaneous sensations). Gastrointestinal disturbances such as loss of appetite, flatulence, diarrhea, constipation, nausea, and vomiting are also reported (Fairbanks and Beutler 1972). When the iron deficiency anemia is severe, chronic changes such as koilonychia (spoon-shaped nails), angular stomatitis (cracks at the corner of the mouth), glossitis (sore tongue), flattening of the lingual papillae, atrophic gastritis (stomach inflammation with atrophy of the mucous membranes), and also bone changes in children may occur (Hoffbrand and Lewis 1981). The bone changes are thought to be related to a hyperactive bone marrow that creates pressure on surrounding bone, thus increasing the width of the marrow space and decreasing the width of the outer table of bone.

Sex-related differences in prehistoric populations

Evidence for anemia in prehistoric populations comes from a paleopathology called porotic hyperostosis. Porotic hyperostosis is present when the normally smooth, dense outer compact bone of the skull and orbit is pitted by small holes of varying size and density (Figure 4.1). Often the middle layer of bone, or diploe, is thicker than normal. The lesions are usually symmetrical in distribution and occur mainly on the orbits (also known as cribra orbitalia), and the frontal, parietal, and occipital bones of the skull vault. The relationship between lesions on the orbits and lesions on the skull vault is not absolutely clear, but they both appear to be similar manifestations of marrow hyperplasia (Stuart-Macadam 1989). It is possible that lesions of the vault indicate more long-term or severe cases of anemia, as clinical data from patients with anemia show that skull changes start earliest in the frontal bone (which includes the orbital roofs or the horizontal plates of the frontal bone) and then progress to the rest of the skull vault (Caffey 1937). Clinical studies also show that

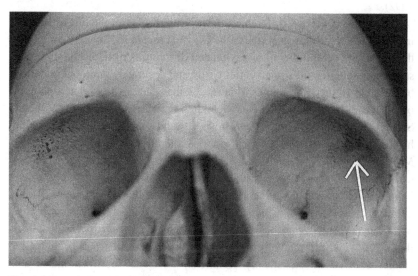

Fig. 4.1 Porotic hyperostosis (also known as cribra orbitalia; arrowed).

the severity of the anemia does not always correspond with the severity of bone change. Radiographs of archaeological skulls with porotic hyperostosis show a pattern of bone changes that includes thinning of the outer table, an increase of the middle table, occasionally a 'hair-on-end' pattern of trabeculation, and thickening of the orbital roof (Stuart-Macadam 1982). Similar skull changes occur in modern clinical patients with various types of anemia (Stuart-Macadam 1982). It is not easy to differentiate the type of anemia on the basis of skeletal changes alone, but a number of factors suggest that most cases of porotic hyperostosis in prehistoric populations were probably associated with iron deficiency anemia (Stuart-Macadam 1982).

Porotic hyperostosis occurs in human populations from almost every time period, continent, and country, and there are temporal, geographic and ecological trends (Stuart-Macadam 1995). Specifically, porotic hyperostosis does not occur in skeletal remains from the Paleolithic era and there are only a few cases in the Mesolithic era (Kennedy 1984; Meiklejohn *et al.* 1984; Grmek 1989). It began to occur more frequently in the Neolithic era, with the adoption of agriculture and the increasing sedentism and aggregation of peoples (Lallo *et al.* 1977; Angel 1978; Cohen and Armelagos 1984; Kent 1986). After the Neolithic era, the frequency varies tremendously in human populations from different time periods and geographic locales, but there is a general decrease in frequency towards the twentieth century and greater frequency in populations closer to the equator, and in lowland compared with highland environments (Henschen 1961; Hengen 1971; Angel 1978; Stuart-Macadam 1992).

Are there sex-related differences in the pattern of iron deficiency anemia in prehistory? Unfortunately, the archaeological record has several limitations that make the determination of sex-related differences difficult. First, at present it is impossible to sex juveniles easily and accurately, although this may be possible in the near future. This means that data on sex-related patterns of anemia in prehistoric infants and children are currently difficult to obtain. Second, lesions of porotic hyperostosis in adults most probably represent healed lesions from childhood episodes of anemia (Stuart-Macadam 1985). Marrow physiology is such that the pressures of an enlarging red marrow can produce bone changes in children but not in adults. Consequently, it will be impossible to obtain data on adult-acquired iron deficiency anemia. Third, although there have been numerous published and unpublished studies on porotic hyperostosis, many of them do not include data on sex. Of 28 studies that do provide information on sex-related differences in cribra orbitalia, 24 studies showed no statistically significant difference between males and females, three showed lesions to be more common in females and one showed

Table 4.1. *Differences by sex in porotic hyperostosis*

Researcher(s)	Year	Location	N	Sexual difference
Orbital lesions				
Brothwell and Browne	1994	England	329	none
Cox (personal communication)	1988	England	229	none
Cybulski	1977	Canada	360	more in females
Eisenberg	1986	USA	395	none
El-Najjar *et al.*	1976	USA	2860	none
El-Najjar *et al.*	1976	USA	336	none
Fairgrieve	1987	Peru	61	none
Grauer	1991, 1993	England	277	none
Grauer	(this volume)	USA	49	none
Hengen	1971	Germany	91	none
Hillson	1980	Nubia	736	none
Hodges	1986	Mexico	119	none
Lanphear	1988	USA	156	more in females
Mays	1991	England	178	none
Mittler and van Gerven	1994	Sudan	26	more in males
Mittler and van Gerven	1994	Sudan	166	none
Møller-Christensen	1953	Denmark	184	none
Nathan and Haas	1966	Israel	514	more in females
Robledo *et al.*	1995	England	221	none
Ryan	1977	USA	389	none
Storey	(this volume)	Honduras	76	none
Stuart-Macadam	1982	England	546	none
Suzuki	1987	Guam	116	none
Suzuki	1987	Hawaii	303	none
Walker	1986	California	397	none
Webb	1982	Australia	400	none
Wells	1982	England	362	none
Wiggins (personal communication)	1993	England	93	none
Vault lesions				
Danforth *et al.*	1997	Belize	185	more in males
Fairgrieve	1987	Peru	60	more in males
Hodges	1986	Mexico	153	none
Lanphear	1988	USA	166	none
Storey	(this volume)	Honduras	104	none
Stuart-Macadam	1982	England	546	more in males

lesions to be more common in males (Table 4.1). Only six studies provided information on sex-related differences in vault lesions: three of these showed lesions to be more common in males and three showed no difference.

Sex-related differences in historic and contemporary populations

The recorded history of iron deficiency anemia is obscure, partly because symptoms are often vague and poorly defined. The Papyrus Ebers, an Egyptian manual of therapeutics from 1500 BC, describes a disease characterized by pallor, dyspnea, and edema (fluid accumulation) that may have been iron deficiency anemia (Fairbanks and Beutler 1972). There is a possibility that iron deficiency anemia was known in fourth-century Britain; a small bronze forearm with fingernails showing koilonychia was found at Lydney Park Temple, a Romano-British site in Gloucestershire, England associated with ancient iron workings (Hart 1981). However, these early references provide no clue about sex-related differences.

In 1554, a disorder called chlorosis (derived from the Greek word for green) was first described; this was probably a type of iron deficiency anemia. It became popularly known as the 'green-sickness', and was depicted in many paintings by the Dutch masters and alluded to by Shakespeare and other writers (Farley and Foland 1990). It commonly affected teenage girls and was thought to be associated with emotional upsets or 'love-sickness'. Chlorosis became common in the late nineteenth century but had virtually disappeared by the early twentieth century. No one knows exactly what caused the condition, although it has been suggested that a combination of the demands of growth, the onset of menses, inadequate diet, and tight corsets may have contributed to it (Lee *et al.* 1993). In the late 1920s and early 1930s another form of iron deficiency anemia was recognized, 'chronic hypochromic' anemia. This time it was women who were more commonly affected, and the anemia was associated with poor diet, multiple pregnancies, and menstrual irregularities.

Today it is known that iron deficiency anemia can affect any sex and age group, although infants, young children, and women in their reproductive years are said to be most vulnerable. What do modern clinical and epidemiological studies show with respect to sex-related differences? Interestingly, although modern studies of iron deficiency anemia in adults always consider sex, few studies of infants and children do. For example, of 13 studies examining the incidence of iron deficiency anemia in infants from birth to two years (Table 4.2), only three looked at sex-related differences. In all three studies (Woodruff 1958; Cathilineau 1964; Betke 1970) iron deficiency anemia was more common in infant boys. In children aged from 2 to 15 years the picture was more ambiguous; where sex-related differences were recorded, sometimes it was more common in boys (Fernandes-Costa *et al.* 1984; Expert Scientific Working Group 1985), sometimes in girls (Florentino and Guirriec 1984;

Table 4.2. *Sex-related differences in iron deficiency anemia from birth to two years*

Researcher(s)	Year	Location	N	Sexual difference
Betke	1970	Germany	?	more in males
Calvo and Gnazzo	1990	Argentina	384	not examined
Cathilineau	1964	France	58	more in males
Davidson *et al.*	1942	Scotland	442	not examined
Davis *et al.*	1960	England	114	not examined
Expert Scientific Working Group	1985	USA	?	not examined
Florentino and Guirriec	1984	Africa	?	not examined
Florentino and Guirriec	1984	Bangladesh	?	not examined
Florentino and Guirriec	1984	Bolivia	?	not examined
Florentino and Guirriec	1984	Brazil	?	not examined
Florentino and Guirriec	1984	Caribbean	?	not examined
Florentino and Guirriec	1984	Chile	?	not examined
Florentino and Guirriec	1984	Indonesia	?	not examined
Florentino and Guirriec	1984	Malaysia	2103	not examined
Fuerth	1971	USA	?	not examined
Guest and Brown	1957	USA	214	not examined
Lehmann *et al.*	1992	Canada	218	no difference
Poppe	1993	New Zealand	?	not examined
Shrestha *et al.*	1994	Fiji	2136	not examined
Woodruff	1958	USA	31,113	more in males

Hercberg 1992), and sometimes there was no difference (Simonivits *et al.* 1970; Vellar 1970; Westhuyzen and Steyn 1992) (Table 4.3). It is difficult to get figures on infants and children because of a lack of national data, and variations in methods and criteria used for diagnosis (Florentino and Guirriec 1984).

Although it would appear that the examination of sex-related differences in adults would be a straightforward process, it is actually quite problematic. There is little standardization with respect to sample populations surveyed, definition of anemia, or even collection and analyses of data, making it very difficult to compare data among studies. As well, there has been little appreciation of normal physiological variation in iron status, especially for women. For example, a pregnant or lactating woman will show lower hemoglobin levels for physiological, not pathological, reasons and yet in most studies reproductive status was not considered. Few studies differentiate between the anemia of infection or chronic disease and iron deficiency anemia, which obscures the true picture of iron deficiency anemia. However, seven studies surveying the prevalence of iron deficiency anemia in more than 10,000 adults all show anemia to be more common in women than men (Table 4.4).

Table 4.3. *Sex-related differences in iron deficiency anemia in children aged from 2 to 15 years*

Researcher(s)	Year	Location	N	Sexual difference
Davidson *et al.*	1942	England	389	not examined
Expert Scientific Working Group	1985	USA	?	more in males (11–14 years)
Fernandes-Costa	1984	South Africa	32	more in males
Florentino and Guirriec	1984	Bangladesh	?	no difference
Florentino and Guirriec	1984	India	1110	not examined
Florentino and Guirriec	1984	China	1148	not examined
Florentino and Guirriec	1984	Philippines	175	not examined
Florentino and Guirriec	1984	Philippines	231	more in females
Hercberg	1992	France	?	more in females (adolescent)
Simonivits *et al.*	1970	Hungary	1039	no difference
Vellar	1970	Norway	?	no difference
Westhuyzen and Steyn	1992	South Africa	?	no difference

Table 4.4. *Sex-related differences in iron deficiency anemia in adults*

Researcher(s)	Year	Location	N	Sexual difference
Hallberg	1970	Sweden	1877	more in females
Jonsson *et al.*	1991	Iceland	4240	more in females
Kilpatrick	1970	England	?	more in females
Kilpatrick	1970	Wales	800	more in females
Kilpatrick	1970	Wales	1800	more in females
MacPhail and Bothwell	1992	Africa		more in females
MacPhail and Bothwell	1992	North America		more in females
MacPhail and Bothwell	1992	Latin America		more in females
MacPhail and Bothwell	1992	East Asia		more in females
MacPhail and Bothwell	1992	South Asia		more in females
MacPhail and Bothwell	1992	Europe		more in females
MacPhail and Bothwell	1992	Oceania		more in females
Seibold	1970	Germany	1216	more in females
Sultan	1964	France	147	more in females
Vellar	1970	Norway	1439	more in females

Summary of past and contemporary human studies

If porotic hyperostosis is largely representative of iron deficiency anemia, then it appears that this type of anemia was non-existent in the Paleolithic, rare in the Mesolithic, and started to become more common in the Neolithic era. The pattern after that time is variable, but anemia seemed to decrease towards the twentieth century, and to occur less frequently with increasing land elevation and decreasing proximity to the equator. The archaeological record has limitations, which make interpretation of the pattern of iron deficiency anemia difficult. If adult lesions of porotic hyperostosis are representative of childhood episodes of iron deficiency anemia, then it appears that there are few, if any, sex-related differences among prehistoric children. The fact that the only studies that examine sex-related differences in skull lesions of porotic hyperostosis show males affected more than females could indicate that in childhood, boys had more severe episodes of iron deficiency anemia. However, this is questionable, since bone changes are not always indicative of clinical severity.

Up until the sixteenth century, historical records provide no information on sex-related differences in iron deficiency anemia. If chlorosis was a form of iron deficiency anemia, then it appears that young women were more frequently affected up until the late nineteenth century. In the late 1920s and early 1930s iron deficiency anemia became common in women during their reproductive years. Modern surveys indicate that today iron deficiency anemia affects mainly infants, children, and premenopausal women. As few studies of young children examine sex-related differences, it is difficult to be sure that they exist, although the three studies that did differentiate the sexes all show that iron deficiency anemia is more common in young boys. The pattern in older children is ambiguous, with some studies showing no sex-related differences, some showing iron deficiency anemia to be more common in girls and some in boys.

The biological anthropologist's view of iron deficiency anemia

With the synthesis of data on iron deficiency anemia from the past and present, and the adoption of a biocultural and cross-cultural perspective, an interesting picture of iron deficiency anemia emerges. It is a condition that appears to be almost non-existent before 10,000 years ago, prior to the beginnings of agriculture and domestication of animals in the Neolithic era. The last 10,000 years actually represent a tiny fraction of our total time on Earth, and in fact 99.9% of human existence has been as hunter–gatherers (Eaton *et al.* 1988).

Iron deficiency anemia, it would appear, is a condition that did not occur in the hunting and gathering phase of human evolution. This is supported by studies of contemporary hunter–gatherers who still follow traditional life-styles; these people are free from iron deficiency anemia (Metz *et al.* 1971), and it is only after they assume a settled lifestyle that anemia becomes a problem (Fernandes-Costa *et al.* 1984; Kent and Lee 1992). This means that, in evolutionary terms, iron deficiency anemia is a very recent phenomenon. The presence of skeletal lesions associated with iron deficiency anemia indicates a childhood condition, but anthropological studies cannot provide data as to how iron deficiency anemia has affected adults. The studies, however, do indicate that iron deficiency anemia was equally common in prehistoric boys and girls, although perhaps more severe in boys. Historical records suggest that iron deficiency anemia most often affected young girls and women. Modern clinical and epidemiological data show that, at present, iron deficiency anemia is very common. It affects women more than men, and possibly infant boys more than infant girls.

Iron metabolism is a complex and poorly understood process; there are a number of factors that can inhibit or enhance iron status including diet, physiology, and genetics (Table 4.5). Although the mechanism is unknown, the intestinal mucosa can regulate the absorption of iron to a certain extent; iron-depleted individuals can increase their absorption of iron from food and iron-replete individuals can decrease their absorption. Children, who have greater requirements for nutrients, including iron, normally absorb much more iron than adults, and women absorb about twice as much as men (Wadsworth 1975). Although it is not widely known, iron is also involved in human immune system function. Many pathogenic microorganisms, including bacteria, require iron for metabolic processes but lack their own stores. However, they do have the ability to manufacture their own iron-binding proteins, siderophores, which can latch onto free serum iron. When exposed to disease-causing microorganisms, the human body goes into a state of iron-withholding, or hypoferremia (Weinberg 1992). This involves stationing iron-binding proteins at sites of entry for pathogens, binding available serum iron onto iron transporting and storing proteins, and reducing the amount of iron absorbed from food (Weinberg 1992). This ensures that in the battle for iron, the pathogenic microorganisms are at a disadvantage. In the same situation, iron-overload or too much free iron in the serum can give the disease-causing microorganisms the advantage.

Iron metabolism is almost a closed system; as much as 90% of the iron needed for incorporation into newly forming red blood cells is obtained by the recycling of old red blood cells that have been destroyed by the

Table 4.5 *Factors involved in iron status*

Iron absorption
- Children absorb more than adults
- Females absorb more than males
- After first trimester pregnant women absorb more than non-pregnant women
- Lactating women absorb more than non-lactating
- Iron-deficient individuals absorb more than iron-replete
- Cultural factors such as use of tobacco, alcohol, iron pots, iron supplements, etc., that increase iron stores

Diet
- Factors inhibiting iron absorption
  - phytates (cereals, nuts, legumes)
  - calcium in the form of milk, cheese, or supplements
  - polyphenols – plant metabolites present in tea, coffee, cocoa, red wine, vegetables, and legumes
- Factors enhancing iron absorption
  - ascorbic acid
  - meat, fish, poultry
  - fermented foods, including miso

Blood loss
- Parasites (hookworm, *Giardia* spp, *Plasmodium* spp, *Schistosoma* spp)
- Gastrointestinal disorders (hemorrhoids, ulcers, ulcerative colitis)
- Drug use (aspirin, antibiotics)
- Cow's milk (in infants)
- Menstruation

Iron withholding
- Hypoferremia as a defense
- Anemia of infection
- Anemia of chronic disease

Genetics
- Hemochromatosis
- Hemoglobinopathies

---

reticuloendothelial system (Hoffbrand and Lewis 1981). This means that very little iron is required from the diet. Apart from menstruation, there are no physiological mechanisms for iron loss; as a result, iron can accumulate in the body to toxic levels. However, as mentioned above, hypoferremia is an important aspect of our immune system. These facts suggest that in our evolutionary past bio-available iron was not readily accessible, therefore it was more critical for the body to conserve iron than to deal with the hazards of excess iron.

For an individual to maintain a healthy iron metabolism there must be a balance between too little iron and too much iron. Too little iron can be responsible for a compromised immune system, impaired work performance and exercise capacity, poor pregnancy outcome, possibly impaired cognitive and motor development in children, and bone changes in children related to hyperplasia of bone marrow (Stuart-Macadam 1996). Too much iron can lead to a compromised immune system, increased risk of Sudden Infant Death Syndrome, increased risk of some cancers and possibly myocardial infarction, increased risk of disease and death in people who are homozygous for hemo-chromatosis (a disorder resulting from an inherited intestinal absorption defect that leads to increased absorption of iron and eventual iron overload), fibrotic scarring, and the failure of several organs, including the liver, pancreas, heart and endocrine system (Stuart-Macadam 1996). It appears that, for much of our time on Earth, the human body has been in balance with respect to iron metabolism, and people did not suffer from iron deficiency anemia. Since the Neolithic era, however, this has changed, and humans have been much more susceptible to this condition. Conversely, many contemporary humans, especially men, are now suffering from the detrimental effects of too much iron. Why is this so?

Culture is a potent factor affecting our biology, just as biological factors put constraints on culture. In the hunting and gathering phase of human history there were thousands of years of continuity in terms of human interaction with the environment. With the domestication of plants and animals in the Neolithic era came a radical change in subsistence activities and the kinds of interactions that existed between humans and their environment. For millions of years an equilibrium had existed in terms of iron metabolism – there was enough iron for metabolic needs but not too much, which would be advantageous to pathogenic microorganisms, or too little, which would result in iron deficiency anemia. In the Neolithic period, all this began to change. Populations were exposed to new and increasing numbers of microorganisms, partly because of the increasing proximity to animals, hence zoonoses, and partly because of increasing sedentism and aggregation of peoples, resulting in poorer hygiene and larger population vectors for disease. Greater pathogen loads would have challenged the immune system, resulting in the iron-withholding mechanism being activated, and parasites such as hookworm and the malarial parasite (*Plasmodium falciparum*) would have contributed to direct blood loss. This increased pathogen load could have upset the physiological iron balance that had been maintained for millennia and taken the adaptive immune system response of hypoferremia over the threshold into an iron deficiency anemia. In this case, both males and females would have been affected by the anemia.

Young girls are known to have higher levels of IgM (immunoglobulin M) than young boys (Purtilo and Sullivan 1979), and perhaps, all things being equal, are better able to overcome infectious diseases than young boys. If the concept of enhanced female immune reactivity is true (see Ortner, Chapter 6) then perhaps it could explain why some prehistoric boys may have had more severe or chronic iron deficiency anemia than girls, and why, even today, it appears that more male babies suffer from iron deficiency anemia than female babies. Until very recently in human history, women breast-fed their babies and children for at least the first two to four years of life (Stuart-Macadam and Dettwyler 1995). The combination of pregnancy and the contraceptive effect of lactational amenorrhea meant that for much of their reproductive lives women did not have menstrual cycles. In fact, Short (1976) says that on average, a woman would have experienced only about 50 menstrual cycles during her reproductive lifetime, as opposed to about 450 for the modern woman (Eaton *et al.* 1994). Consequently, women in the past did not normally have monthly menstrual cycles and the blood losses (hence iron losses) associated with them (see Weaver, Chapter 3, for discussion of the effects of increased menstrual cycles on osteoporosis). This began to change in recent history, particularly since the Medieval period in Europe, when wealthy and upper class women began to use wet nurses to feed their babies. This meant that for the first time in millions of years of human evolution, some women began to have uninterrupted regular menstrual cycles, and, without the contraceptive effect of lactation, repeated and numerous pregnancies. If the !Kung peoples of the Kalahari desert could be used as an analogy for our hunting and gathering ancestors (although this must be done with caution), it would suggest that the average woman in prehistory had a child about every four years, totaling only five or six children during her reproductive lifetime (Lee and DeVore 1976). This contrasts with upper class women in seventeenth century Europe, for example, who were encouraged to have large families and hired wet nurses for their infants; 15–20 children were not uncommon (McLaren 1978). The metabolic stresses of repeated pregnancies, childbearing, and the blood losses of repeated menstrual cycling between pregnancies no doubt contributed to iron deficiency anemia. So, from at least medieval times, women would have been more susceptible to iron deficiency anemia than men.

Breast-fed babies rarely suffer from iron deficiency anemia, but it commonly occurs in infants fed cow's milk or unfortified cow's milk-based infant formula (Woodruff 1958; Calvo and Gnazzo 1990; Lehmann *et al.* 1992; Shrestha *et al.* 1994). Several factors contribute to the higher incidence of iron deficiency anemia in formula-fed babies. These include the reduced bioavailability of iron in cow's milk, gastrointestinal bleeding as the result of intolerance to cow's milk or cow's milk-based infant formula, and a 50% reduction in absorption of both heme and nonheme iron in other foods resulting from the high levels of calcium occurring in cow's milk (Stuart-Macadam 1995). In the western world, infant formula began to be used on a widespread basis from about the 1950s, and its use has now spread around the world. This, among other factors, has contributed to an enormous reduction in contemporary society in the numbers of mothers who breast-feed and of babies who are breast-fed. Up until very recently in human history, all babies were breast-fed; if they weren't, they died. Prehistoric infants rarely show signs of iron deficiency anemia under the age of one year (Stuart-Macadam 1982). It is very likely that the greater frequency of iron deficiency anemia occurring in modern infants is partly related to the fact that they are not receiving human breastmilk. In the contemporary world some societies are still exposed to heavy pathogen loads and poor nutrition, with large numbers experiencing iron deficiency anemia. At the same time, other societies are exposed to excessive amounts of iron as the result of heavily fortified foods, contaminated water, or customs that result in the incorporation of large amounts of iron into their bodies (Kent *et al.* 1990). In the latter case some people, particularly men, women after menopause, and those individuals with the gene for hemochromatosis, are at risk for disease and possibly death from too much iron. Recent studies have shown a link between high iron levels and both heart disease and cancer, and hemochromatosis can be fatal if not treated (Stuart-Macadam 1996). Both iron deficiency anemia and iron over-load compromise the immune system, resulting in increased susceptibility to infection.

Conclusion

It appears that, until very recently in human history, the body was in equilib-rium in terms of iron physiology. There were no sex-related differences in iron deficiency anemia because, as is the case with contemporary hunters and gatherers, the condition didn't exist. That began to change about 10,000 years ago. Iron deficiency anemia appeared when a settled way of life, with increased exposure to disease-causing microorganisms, began to replace a nomadic, hunting and gathering existence. The limitations of archaeological data mean that it is not possible to determine the pattern of iron deficiency anemia for prehistoric men and women, however, the data show that there were no differences in frequency between boys and girls, although anemia may have been more severe in boys. Sex-related differences in the frequency of the condition were not apparent until medieval times, when iron deficiency anemia occurred more commonly in young females. Later in history it assumed the modern pattern, with females in their reproductive years, infants, and children being most commonly affected. Bioavailable iron was probably not readily obtainable in our evolutionary past, hence the fact that there are no physiological mechanisms for the elimination of iron. This has largely protected men from iron deficiency anemia (although in modern times they are now vulnerable to the dangers of excess iron), but women have been more susceptible because of the unusual iron losses (in evolutionary terms) associated with repeated menstrual cycling and/or numerous pregnancies, and babies and children because of the lack of breast-feeding along with disease and the extra demands of growth and development. The few modern clinical studies that examine differences between infant boys and girls suggest that iron deficiency anemia is more common in infant boys, perhaps related to their weaker immune systems.

The application of the biocultural, evolutionary, and cross-cultural approach of the biological anthropologist illustrates that cultural factors, such as changing subsistence strategies or infant feeding styles, can have a profound affect on our biology. As a result, iron deficiency anemia, a condition that did not even exist for most of human evolution, currently affects millions of people around the globe. Contrary to what many people believe, iron deficiency anemia is not just a dietary deficiency of iron that needs to be corrected by iron supplementation. To the biological anthropologist it is a condition that is bound to individual physiology and immune system function, and embedded within a temporal, ecological, social, economic, and political framework. Its changing prevalence throughout time and space reflects changing relationships between humans and their environment. Sex-related differences in iron deficiency anemia provide clues about human ecology in the past and present. Applying the perspective of the biological anthropologist reveals the complexity of the story of iron deficiency anemia and illustrates the powerful affect that culture can have on human biology.

References

Angel JL (1978) Porotic hyperostosis in the Eastern Mediterranean. *Medical College of Virginia Quarterly* **14**(1):10–16.

Betke K (1970) Iron deficiency in children. In L Hallberg, HG Harwerth, and A Vannotti (eds.), *Iron Deficiency*. London: Academic Press, pp. 519–24.

Bridges KR and Seligman PA (1995) Disorders of iron metabolism. In RI Handin, SE Lux, and TP Stosell (eds.), *Blood: Principles and Practice of Hematology*. Philadelphia: JB Lippincott Company, pp. 1433–72.

Brothwell DR and Browne S (1994) Pathology. In JM Lilley, G Stroud, DR

Brothwell and MH Williamson (eds.), *The Jewish Burial Ground at Jewbury.*
The Archaeology of York, Volume 12(3): The Medieval Cemeteries. York:
Council for British Archaeology, pp. 457–94.

Caffey J (1937) The skeletal changes in the chronic hemolytic anemias. *American
Journal of Roentgenography and Radiation Therapy* **65**(4):547–60.

Calvo EB and Gnazzo N (1990) Prevalence of iron deficiency in children aged
9–24 mo from a large urban area of Argentina. *American Journal of Clinical
Nutrition* **52**:534–40.

Cathilineau L (1964) Quoted in Dresch C (1970) Prevalence of iron deficiency in
France. In L Hallberg, HG Harwerth, and A Vannotti (eds.), *Iron Deficiency.*
London: Academic Press, pp. 423–5.

Cohen MN and Armelagos GJ (Eds.) (1984) *Paleopathology at the Origins of
Agriculture.* New York: Academic Press.

Cybulski JS (1977) Cribra orbitalia, a possible sign of early historic native
populations of the British Columbia coast. *American Journal of Physical
Anthropology* **47**:31–40.

Danforth ME, Jacobi KP, and Cohen MN (1997) Gender and health among the
colonial Maya of Tipu, Belize. *Ancient Mesoamerica* **8**:13–22.

Davidson LSP, Donaldson GMM, Dyar MJ, Lindsay ST, and McSorby JG (1942)
Nutritional iron deficiency in wartime. *British Medical Journal* No.10/31:
505–7.

Davis LR, Marten RH, and Sarkany I (1960) Iron deficiency anemia in European
and West Indian infants in London. *British Medical Journal* No.11/12:
1426–8.

Early R and Wotecki E (1993) *Iron Deficiency Anemia: Recommended
Guidelines for the Prevention, Detection, and Management Among U.S.
Children and Women of Childbearing Age.* Washington: National Academy
Press.

Eaton SB, Pike MC, Short RV, Les NC, Trussell J, Hatcher RA, Wood JW,
Worthman CM, Blurton Jones NG, Konner MJ, Hill KR, Bailey R, and
Hurtado AM (1994) Women's reproductive cancers in evolutionary context.
Quarterly Review of Biology **69**:353–67.

Eaton SB, Shostak M, and Konner M (1988) *The Paleolithic Prescription.* New
York: Harper and Row Publishers.

Eisenberg L (1986) *Adaptation in a 'Marginal' Mississippian Population from
Middle Tennessee: Biocultural Insights From Paleopathology.* PhD
dissertation, New York University.

El-Najjar MY, Ryan DJ, Turner CG, and Lozoff B (1976) The etiology of porotic
hyperostosis among the prehistoric and historic Anasazi Indians of
Southwestern United States. *American Journal of Physical Anthropology*
47:447–88.

Expert Scientific Working Group (1985) Summary of a report on assessment of the
iron nutritional status of the United States population. *American Journal of
Clinical Nutrition* **42**:1318–30.

Fairbanks VF and Beutler E (1972) Erythrocyte disorders. In W Williams (ed.),
Haematology. New York: McGraw Hill, pp. 305–19.

Fairgrieve S (1987) *The Pasamayo Crania of the Hutchinson Collection: A
Pathological and Nutritional Assessment.* MPhil dissertation, University of
Cambridge.

Farley P and Foland J (1990) Iron deficiency anemia: How to diagnose and correct.
Postgraduate Medicine **87**(2):89–101.

Fernandes-Costa FJ, Marshall J, Ritchie C, van Tonder SV, Dunn D, Jenkins T, and Metz J (1984) Transition from a hunter–gatherer to a settled lifestyle in the !Kung San: effect on iron, folate, and vitamin B12 nutrition. *American Journal of Clinical Nutrition* **40**:1295–303

Florentino R and Guirriec R (1984) Prevalence of nutritional anemia in infancy and childhood with emphasis on developing countries. In A Stekel (ed.), *Iron Nutrition in Infancy and Childhood.* New York: Vevey/Raven Press, pp. 61–74.

Fuerth J (1971) Incidence of anemia in full-term infants seen in private pratice. *Journal of Pediatrics* **79**(4):560–2.

Grauer AL (1991) Life patterns of women from Medieval York. In D Walde and ND Willows (eds.), *The Archaeology of Gender.* Proceedings of the Twenty-Second Annual Conference of the Archaeological Association of the University of Calgary. Calgary: University of Calgary Archaeological Association.

Grauer AL (1993) Patterns of anemia and infection from Medieval York, England. *American Journal of Physical Anthropology* **91**(2):203–13.

Grmek MD (1989) *Diseases in the Ancient Greek World.* (Translated by M Muellner and L Muellner) Baltimore: Johns Hopkins University Press.

Guest GM and Brown EW (1957) Erythrocytes and hemoglobin of the blood in infants and children. *American Journal of Diseases of Children* **93**:486–509.

Hallberg L (1970) Prevalence of iron deficiency in Sweden. In L Hallberg, HG Harwerth, and A Vannotti (eds.), *Iron Deficiency.* London: Academic Press, pp. 453–9.

Hart GD (1981) Anemia in ancient times. *Blood Cells* **7**:485-9.

Hengen OP (1971) Cribra orbitalia, pathogenesis and probable etiology. *Homo* **22**(2):57–75.

Henschen F (1961) Cribra cranii – a skull condition said to be of racial or geographical nature. *Pathologia et Microbiologia* **24**:724–9.

Hercberg (1992) Comment. Quoted in P Dallman (1992) Changing iron needs from birth through adolescence. In SJ Fomon and S Zlotkin (eds.), *Nutritional Anemias.* Nestle Nutrition Workshop Series, volume 30. New York: Vevey/Raven Press, pp. 29–38.

Hillson S (1980) *Human Biological Variation in the Nile Valley in Relation to Environmental Factors.* PhD dissertation, University of London.

Hodges D (1986) *Agricultural Intensification and Prehistoric Health in the Valley of Mexico.* PhD dissertation, State University of New York, Albany.

Hoffbrand AV and Lewis SM (1981) *Postgraduate Haematology.* London: William Heinemann Medical Books Limited.

Jonsson JJ, Johannessun GM, Sigfusson N, Magnusson B, Thjodleifsson B, and Magnusson S (1991) Prevalence of iron deficiency and iron overload in the adult Icelandic population. *Journal of Epidemiology* **44**:1289-97.

Kennedy KAR (1984) Trauma and disease in the ancient Harappans. In BB Lal and SP Gupta (eds.), *Frontiers of the Indus Civilization.* New Delhi: Books and Books, pp. 425–36.

Kent S (1986) The influence of sedentism and aggregation on porotic hyperostosis and anemia: A case study. *Man* **21**:605–36.

Kent S and Lee R (1992) A hematological study of !Kung Kalahari foragers: an eighteen-year comparison. In P Stuart-Macadam and S Kent (eds.), *Diet, Demography, and Disease: Changing Perspectives on Disease.* New York: Aldine de Gruyter, pp. 173–200.

Kent S, Weinberg ED, and Stuart-Macadam P (1990) Dietary and prophylactic iron supplements: helpful or harmful. *Human Nature* **1**: 53–79.

Kilpatrick GS (1970) Prevalence of anemia in the United Kingdom. In L Hallberg, HG Harwerth, and A Vannotti (eds.), *Iron Deficiency*. London: Academic Press, pp. 441–5.

Lallo JW, Armelagos GJ, and Mensforth RP (1977) The role of diet, disease and physiology in the origin of porotic hyperostosis. *Human Biology* **49**(3):471–83.

Lanphear K (1988) *Health and Mortality in a Nineteenth Century Poorhouse Skeletal Sample.* PhD dissertation, State University of New York, Albany.

Lee RB and DeVore (Eds.) (1976) *Kalahari Hunter-Gatherers.* Cambridge, Mass: Harvard University Press.

Lee RG (1993) Iron deficiency and iron deficiency anemia. In GR Lee, TC Bithell, J Foerster, JW Athens, and JN Lukens (eds.), *Wintrobe's Clinical Hematology*, volume 1, 9th edn. Philadelphia: Lea and Febiger, pp. 808–39.

Lehmann FK, Gray-Donald K, Mongeon M, and Di Tommaso (1992) Iron deficiency anemia in 1-year-old children of disadvantaged families in Montreal. *Canadian Medical Association Journal* **146**(9):1571–7.

Lux SE (1995) Introduction to anemias. In RI Handin, SE Lux, and TP Stossel (eds.), *Blood: Principles and Practice of Hematology*. Philadelphia: JB Lippincott Company, pp. 1383–97.

MacPhail P and Bothwell TH (1992) The prevalence and causes of nutritional iron deficiency anemia. In S Fomon and S Zlotkin (eds.), *Nutritional Anemias*. Nestle Nutrition Workshop Series, volume 30. New York: Vevey/Raven Press Ltd., pp. 1–12.

Mays S (1991) *The Burials From the Whitefriars Friary Site, Buttermaket, Ipswich, Suffolk (Excavated 1986–1988).* Ancient Monument Laboratory Report Series 17/91. London: English Heritage.

McLaren D (1978) Fertility, infant mortality and breast-feeding in the 17th century. *Medical History* **22**:378-96.

Meiklejohn C, Schentag C, Venema A, and Key P (1984) Socioeconomic change and patterns of pathology in the Mesolithic and Neolithic of Western Europe. Some suggestions. In MN Cohen and GJ Armelagos (eds.), *Paleopathology at the Origins of Agriculture*. New York: Academic Press, pp. 75–100.

Metz J, Hart D, and Harpending HC (1971) Iron, folate, and vitamin B12 nutrition in a hunter–gatherer people: a study of the !Kung Bushmen. *American Journal of Clinical Nutrition* **24**:229–42.

Mittler DM and van Gerven DP (1994) Developmental, diachronic, and demographic analysis of cribra orbitalia in the Medieval Christian populations of Kulubnarti. *American Journal of Physical Anthropology* **93**:287–97.

Møller-Christensen V (1953) *Ten Lepers from Naestved in Denmark*. Copenhagen: Danish Science Press.

Nathan H and Haas N (1966) 'Cribra orbitalia' A bone condition of the orbit of unknown nature. *Israel Journal of Medical Sciences* **2**:171–91.

Poppe M (1993) Iron deficient children. *New Zealand Medical Journal* **4**:392.

Purtilo DT and Sullivan J (1979) Immunological bases for superior survival of females. *American Journal of Diseases of Children* **133**:1251–3.

Robledo B, GJ Trancho, and Brothwell D (1995) Cribra orbitalia: health indicator in the late Roman population of Cannington (Somerset, Great Britain). *Journal of Palaeopathology* **7**(3):185–93.

Ryan DJ (1977) *The Paleopathology and Paleoepidemiology of the Kayenta Anasazi Indians in Northeastern Arizona.* PhD dissertation, Arizona State University.

Seibold M (1970) Prevalence of iron deficiency in Germany. In L Hallberg, HG

Harwerth, and A Vannotti (eds.), *Iron Deficiency*. London: Academic Press, pp. 427–39.

Short RV (1976) The evolution of human reproduction. *Proceedings of the Royal Society, London* **B195**:3–24.

Shrestha M, Chandra V, and Singh P (1994) Severe iron deficiency in Fiji children. *New Zealand Medical Journal* **107**:130–2.

Simonivits I, Lepes P, Simon T, Budai B, Bodnar L, and Szasz D (1970) Studies of the epidemiology of anemia. *Transfusio* **4**:4–14.

Stuart-Macadam PL (1982) *A Correlative Study of a Palaeopathology of the Skull*. PhD dissertation, Department of Physical Anthropology, University of Cambridge.

Stuart-Macadam PL (1985) Porotic hyperostosis: representative of a childhood condition. *American Journal of Physical Anthropology* **66**:391–8.

Stuart-Macadam PL (1989) Porotic hyperostosis: relationship between orbital and vault lesions. *American Journal of Physical Anthropology* **74**(4):511–20.

Stuart-Macadam PL (1992) Porotic hyperostosis: a new perspective. *American Journal of Physical Anthropology* **87**:39–47.

Stuart-Macadam PL (1995) Biocultural perspectives on breast-feeding. In P Stuart-Macadam and K Dettwyler (eds.), *Breast-feeding: Biocultural Perspectives*. New York: Aldine de Gruyter, pp. 1–38.

Stuart-Macadam PL (1996) Paleopathology does have relevance to contemporary issues. In A Perez-Perez (ed.), *Notes on Populational Significance of Paleopathological Conditions: Health, Illness and Death in the Past*. Barcelona: Ramargraf, S.A., pp. 123–35.

Stuart-Macadam PL and Dettwyler K (1995) *Breast-feeding: Biocultural Approaches*. New York: Aldine de Gruyter Press.

Sultan Y (1964) Quoted in Dresch C (1970) Prevalence of iron deficiency in France. In L Hallberg, HG Harwerth, and A Vannotti (eds.), *Iron Deficiency*. London: Academic Press, pp. 423–25.

Suzuki T (1987) Cribra orbitalia in the early Hawaiians and Mariana Islanders. *Man and Culture in Oceania* **3**:95–104.

Vellar OD (1970) Prevalence of iron deficiency in Norway. In L Hallberg, HG Harwerth, and A Vannotti (eds.), *Iron Deficiency*. London: Academic Press, pp. 447–52.

Wadsworth GR (1975) Nutritional factors in anemia. *World Review of Nutrition and Dietetics* **21**:75–150.

Walker P (1986) Porotic hyperostosis in a marine-dependent California Indian population. *American Journal of Physical Anthropology* **69**:345–54.

Webb S (1982) Cribra orbitalia: a possible sign of anemia in pre-and post-contact crania from Australia and Papua New Guinea. *Archaeology and Physical Anthropology in Oceania* **17**:148–56.

Weinberg E (1992) Iron withholding in prevention of disease. In P Stuart-Macadam and S Kent (eds.), *Demography, and Diet, Disease: Changing Perspectives on Anemia*. New York: Aldine de Gruyter, pp 105–50.

Wells C (1982) The human burials. In A Whirr, L Viner, and C Wells (eds.), *Romano-British Cemeteries at Cirencester*. Cirencester: Cirencester Excavations Committee, Corinium Museum.

Westhuyzen J and Steyn N (1992) Hematological nutrition of school children in the Far Northwestern Cape. *Tropical Geographic Medicine* **44**:47–51.

Wintrobe M (1993) *Clinical Hematology*. Philadelphia: Lea and Febiger.

Woodruff C (1958) Multiple causes of iron deficiency in infants. *Journal of the American Medical Association* **167**:715–20.

5

Sex differences in trace elements:
status or self-selection?

DELLA COLLINS COOK and KEVIN D. HUNT

The emerging field of paleonutrition is a small, highly specialized, and extremely active area of research within anthropology. Evolutionary biology teaches us that change can be rapid in small populations. This seems to be as true for the acquisition of new explanatory models among small populations of anthropologists as it is for physical evolution among the populations we study! Because of the field's small size, the application of new explanatory models within paleonutrition appear to occur at a sweepstakes tempo. This is, to some extent, characteristic of all fields that seek to explore the skeletal biology of ancient populations. Each of the small interest groups within our field interacts intensely at professional meetings. After an attractive new model is introduced, we go home to our labs and collections, and begin applying the new model. What results is a cascade of publications on topics such as ancient iron deficiency anemia, weaning-related hypoplasia, or tuberculosis. Each new research vogue replaces the last in our journals.

Research into paleonutrition, however, is situated within a larger interpretive context than anthropology alone. The field responds to topical concerns brought about by medical, biochemical and sociological research. Thus, as a society and as scientists, we find ourselves directly interested in issues concerning food and nutrition. A benchmark for the emergence of the field of paleonutrition was the publication by Wing and Brown in 1979. Since its publication, successive vogues, or emphases, in which one aspect of nutrition is stressed over others have been seen. For instance, an initial research focus relied heavily upon the hunting hypothesis in paleoanthropology. In this paradigm, meat intake was seen as the key element in human adaptation and in the development of human social differentiation (Cartmill 1993). The hypothesis also became the central paradigm in paleonutrition (e.g., Wing and Brown 1979; Ambrose 1993; but see Radosevich 1993:296 for a critique). Models stressing calories, such as Reidhead's (1981), or my own on protein-calorie malnutrition (Cook 1979), reflect the priority of these aspects of nutrition in studies of human biology and international development in the Third World. A more recent paradigmatic shift has been proposed by Speth (1989), who views lipids as limiting nutrients. It is tempting to relate this model to our current public health focus on the over-consumption of fats in the diet of the developed world, and to studies measuring the effects of lipid under-consumption in the under-developed world. While 'over-consumption' is being linked to a myriad of health problems, 'under-consumption' has been argued to play a key role in patterns of human reproduction, as is suggested in the 'critical fat hypothesis' (Lancaster 1984).

Hence, it is not surprising that the first models within the field of paleonutrition sought to quantify the proportion of meat to vegetable foods in prehistoric diets through the use of trace element analysis. The presence and variable ratios between strontium (Sr) and calcium (Ca) retained within skeletal tissue became the focus of research into diets of the past. Strontium quantities in living organisms vary inversely to the organism's position on the trophic pyramid. Plants, for instance, display higher levels of strontium than animals. Hence, it can be predicted theoretically that herbivores, by the nature of their diet, will display higher strontium levels than carnivores. As a result, the flesh of herbivores is lower in strontium than the plant foods in their environment, but higher in strontium than the flesh of the predators who prey upon them. Subsequently, strontium was argued to display a linear relationship to meat intake (see Sillen and Kavanagh 1982; Klepinger 1984 for reviews). The relationship between low strontium and carnivorous diets, however, is more complicated than these early studies assumed. While fine-tuned techniques have targeted specific foods, such as nuts, molluscs, and clay minerals, as specific sources of strontium (Schoeninger and Peebles 1981; Katzenberg 1984; Blakely 1989; Burton and Price 1990), and multi-element techniques have become common tools to assess diet (Buikstra *et al.* 1989; Price 1989; Sanford 1993), Burton and Wright (1995) now argue that the relationship between Sr/Ca and diet is non-linear. They assert that foods consumed in small quantities, such as cheese, bony fish, and ashes used as seasoning, can have disproportionate effects on trace element content. This development may destroy any potential that Sr/Ca trace element analysis offers for reconstructing prehistoric diet. It certainly requires us to re-evaluate the means by which we reconstruct dietary patterns.

Trace element and gender inequality in ancient humans

Studies of trace element levels in ancient human bone often report the presence
of sex differences. When sex differences are found, authors regularly use
gender inequality models to interpret the results. Arguments frequently
revolve around the inferior status of women, their unequal access to food,
the sexual division of labor in food production and/or food preparation, and
even the role of etiquette in food consumption (Table 5.1). Importantly,
however, *post hoc* phrases alluding to gender inequality as explanations of
significant differences between the sexes are common, but often heavily quali-
fied. Contributors to this literature have cautioned that there may be little
foundation to their inferences. They are wary of the effects that sex differences
in mineral metabolism and bone physiology may have on trace element analysis (Katzenberg 1984; Klepinger 1984; White and Schwarcz 1989). Indeed,
isotope fractionation studies, which measure major constituents of the diet
more directly than trace element studies, have often concluded that no age
and sex differences can be found in diets of past populations (Lovell *et al.*
1986; White and Schwarcz 1989; Ubelaker *et al.* 1995). The lack of parallel
claims for gender inequality in these studies suggests that status differences
did not uniformly affect the diet of our prehistoric/historic ancestors.

Can we infer gender inequality in ancient societies on the basis of sex
differences in bone trace elements? Blakely (1989) has provided the most
extensive critique of this model using samples from the protohistoric southeast
United States. He points out that greater calcium uptake and excretion during
lactation, perhaps augmented by geophagy and food prohibitions, account for
higher strontium levels in women of reproductive age. The issue of geophagy
is important. We are not accustomed to classifying the use of lime or ashes
in processing maize as geophagy, but these means of food preparation can
provide a significant source of calcium. They were common in the past, and
continue to be widely practiced among New World maize-dependent groups
(Katz *et al.* 1974). Extending Blakely's (1989) cautionary note concerning
the use of Sr/Ca ratios to infer gender differences, Radosevich (1993) has
demonstrated astonishing sex differences in bone strontium in ancient
Harappans. While he attributes the differences to differential diagenesis
between males and females, one wonders whether geophagy was present in
ancient Harappa. This would certainly influence the conclusions.

White and Schwarcz (1989) suggest physiological, rather than nutritional,
causes for sex differences in trace elements. Maya females, they assert, display
lower levels of zinc than males. Low zinc levels have been associated with
low meat diets. However, the analysis of nitrogen stable isotope values in

Table 5.1. *Some statements about Eastern Woodland sex differences in trace elements and stable isotopes, and some cautions*

Claims

- 'The Sr [strontium] differences appear to derive primarily from the males ... Inherent physiological differences between sexes should have resulted in the same effects for both [sites]. The observation of changes between sites ... suggests that the sexual dimorphism observed for Sr ... is the result of environmental factors such as diet, rather than the inherent physiological differences between sexes.' (Lambert *et al.* 1982:292–3.)
- 'Within the Ledders population it is apparent that males and females consumed different foods ... females have generally lower concentrations of zinc and higher levels of strontium than males. Such a pattern might be expected if the males were hunting and consuming a greater proportion of meat while the females were gathering plant foods and more heavily dependent on these sources for their nutritional needs.' (Beck 1985:499.)
- 'The females showing more strontium ... Perhaps as Sillen suggests ... the pattern reflects the elevated reproduction in this rapidly expanding population. On the other hand, this may indicate differential consumption of animal protein between the sexes. It would not be unexpected for males in an agricultural society to have greater access to animal protein, perhaps as a result of opportunistic snacking while hunting.' (Buikstra 1984:228–9.)
- 'Differences in diet within groups appear to have been based on gender. Males ... had significantly elevated levels of Zn [zinc] compared to females, who had elevated Sr levels. Thus females are inferred to have had less access to animal protein than males.' (Buikstra *et al.* 1987:68.)
- 'Because nuts are rich in strontium, this alone could account for the difference ... It appears that during the Middle-late Woodland the subsistence tasks of men changed little, while those of women shifted from gathering to growing food.' (Katzenberg 1984:113–4.)
- 'Domesticated plants are beginning to provide important components of the diet by this time, and perhaps this is reflected in the differences between males and females.' (Price *et al.* 1986:372.)

Cautions

- 'Any correlations with specific dietary components are at present hypothetical, since a firm biological linkage between diet and bone composition is lacking for any element, including Sr.' (Lambert *et al.* 1982:292–3.)
- 'Of all the anthropological applications of strontium determination, nothing is more fraught with potential pitfalls than the determination of status differences. This is particularly perilous when males and females are being compared.' (Klepinger 1984:79.)
- 'although we found age/sex differences in the metabolism of Zn-Cd [zinc-cadmium] and Pb [lead], our analyses did not uncover indications of differential buffering, accumulation or loss in the cases of Ba [barium] and Al [aluminum]. Whether these indications are not to be expected, or are hidden by pecularities of the data is unclear.' (Buikstra *et al.* 1989:209.)
- 'Paleodietary reconstruction based solely on trace elements – without macroscopic or microscopic corroboration from bone, teeth, hair, or mummified tissues – remains a risky business.; (Blakely 1989:182.)
- Thus, it is somewhat surprising that not all of the prehistoric sites show higher female strontium levels. Male/female difference in bone strontium may reflect differences in birth rates among these groups, but such a suggestion remains to be tested.' (Price *et al.* 1986:372.)

these females does not support the conclusion that female diets consisted of lower meat consumption. Therefore, zinc loss in pregnancy and lactation is a possible explanation (Adair 1987). At least two other studies have shown zinc levels to be lower in women than in men (Runia 1987; Buikstra *et al.* 1989). Price and coworkers (1986) present a convincing argument that sex differences in strontium levels may also be the result of reproductive physiology rather than dietary differences. In an experimental study in which rats were fed controlled diets, they showed that bone strontium levels increased during pregnancy and lactation. Sex differences in archaeological populations were then reexamined with this factor in mind.

It is important, therefore, to examine the nutritional needs and metabolism of pregnant and lactating women before concluding that different trace element levels found between females and males in archaeological populations reflect patterns of gender inequality. This is especially true since female hunter–gatherers spend much of their life either pregnant or lactating (Lancaster 1984). Nutritional needs of women are likely to be quite different from those of males. Here studies of nonhuman primates can provide a convenient model, since their diet is presumably unaffected by food taboos and other cultural influences on food selection (Wrangham *et al.* 1994 notwithstanding).

Dietary studies of nonhuman primates

A review of nonhuman primate studies reveals a striking regularity in sex differences in food choice. It is worth remarking, though, how rarely sex differences are reported. Although it is clear that social rank and body size influence diet – and therefore are partly responsible for male-female diet difference (Hunt *et al.*, unpublished data) – there is remarkable consistency across all social systems, across levels of sexual dimorphism, and across species where females wield quite different social power. Where sex differences were noted, males consistently ate more fruit, and females ate more leaf and/or invertebrates (Table 5.2). Sex differences in the consumption of meat, invertebrates, blossoms, and bark, are obvious but less consistent. Orang-utans (*Pongo pygmaeus*) constitute one exception, since their diets do not exhibit a clear pattern of sex differences. One study of orang-utans showed females ate more leaves and bark (Galdikas and Teleki 1981), but another found that females ate more fruit than males (Rodman 1977). No sex differences have been found in bonobo (*Pan paniscus*) diets (Badrian and Badrian 1984). Interestingly, there is evidence that females increase protein consumption during pregnancy, suggesting females have higher protein demands because of reproductive needs. Among three guenon species (*Cercopithecus* spp.), females ate a higher proportion of protein-rich leaves when pregnant (Gautier-Hion 1980). Captive galagos (*Galago senegalensis braccatus*) also consume more protein during pregnancy (Sauther and Nash 1987).

The pattern of higher fruit consumption among males and higher leaf consumption among females is common. This pattern appears consistent regardless of whether females are dominant or subordinate, and regardless of their size in comparison to males. This suggests that reproductive physiology may be the principle explanation for the differences.

Interestingly, 'soil', 'earth', or 'minerals' are rarely mentioned as components of nonhuman primate diets (e.g., in references in Table 5.2). However, among the semi-terrestrial chimpanzees geophagy is not uncommon. Chimpanzees consume ashes in human cooking grates, soils that are rich in sodium chloride (Goodall 1986), and clay soil from termite mounds (Wrangham 1977; Goodall 1986; Mahaney *et al.* 1997). Termite mound clay is rich in potassium, magnesium, and calcium, and has traces of copper, manganese, zinc, and sodium. 'Earth' was recorded as making up between 0.59 and 3.85% of the stomach contents of three sympatric guenons (*Cercopithecus cephus*: 3.85%, *N*=62; *C. nictitans*: 2.14%, *N*=100; *C. pogonias*: 0.59%, *N*=52; Gautier-Hion 1980). Unfortunately, sex differences in geophagy were not reported among these or other nonhuman primates that we know of, an oversight in urgent need of correction.

Dietary studies of modern humans

Reviewing studies of human nutrition is complex. Cultural factors relating to food, and methodological issues in the reporting of dietary data, complicate the analysis of human diet. Nutritionists tend to focus on proximate analyses. That is, they focus on the nutrients consumed rather than on the foods eaten. Moreover, just as we found in dietary studies of nonhuman primates, data on sex differences are rarely reported.

Regardless of these problems, a dietary pattern in humans, broadly parallel to that in nonhuman primates, can be found. Meat, it appears, is often a smaller component in the diet of females than of males (Rosenberg 1980; White 1985; Hurtado and Hill 1990). This pattern appears in agriculturally based and hunter–gatherer based subsistence strategies. Hence, the importance of vegetable foods in the diets of females, especially in hunter–gatherer societies, may be a consequence of sexual division of labor. Other gender-based dietary differences in humans that we encountered in the nutrition literature were expressed in terms of 'food prohibitions' or other stereotypical behaviors, and are anecdotal. For example, Bangali girls are taught from the

Table 5.2. *Dominance, diet and sexual dimorphism*

Species	Sex dominant during feeding	Sex differences in diet	Sexual di-morphism[a]	Social system[b]	Reference
Indri indri	female	female: leaf / male: fruit	0.86	monogamy	Richard (1986); Pollock (1977, 1979)
Homo sapiens	male	female: plants / male: meat	0.84	community	Lancaster (1984)
Cercopithecus aethiops sabaeus	males	female: leaf / male: fruit, flowers	0.75	large m.m. f.b.	Harrison (1983); Badrian and Badrian (1984)
Pan paniscus	equal?	male: meat			van Lawick-Goodall (1968); Goodall (1986)
Pan troglodytes	males	male: fruit, meat	0.75	community	
Cercopithecus cephus		females: leaf, inverts / males: fruit	0.71	unimale	Gautier-Hion (1980)
Cebus capucinus	male?	male: low handling-time inverts, caterpillars, meat	0.71	unimale	Rose (1994)
Cercocebus albigena		females: inverts / male: fruit, bark	0.71	large m.m.	Waser (1977)
Papio spp.	male	female: fruit / male: meat	0.5–0.75	large m.m. f.b.	Harding (1973); Strum (1975); Harding and Strum (1976); Norton (1987); Barton et al. (1993)
Cercopithecus pogonias		female: leaf, inverts / male: fruit	0.67	unimale	Gautier-Hion (1980)
Cercopithecus nictitans	?	female: leaf, inverts / males: fruit	0.64	unimale	Gautier-Hion (1980)
Gorilla gorilla beringei	male	female: stems, wood / male: fruit, thistles, nettles	0.58	f. choice	Fossey and Harcourt (1977); Watts (1985)

Gorilla gorilla gorilla	male	female: leaf male: fruit	0.58	f. choice	Remis (1995)
Colobus badius	males?	male: fruit	0.55	large f. tran	Clutton-Brock (1974, 1977)
Pongo pygmaeus	male	female: leaf, bark	0.54	solitary	Galdikas and Teleki (1981)
Pongo pygmaeus	male	female: fruit, inverts male: bark	0.54	solitary	Rodman (1977)
Macaca sinica	male	female: leaf, inverts male: fruit, seeds	0.52	m.m. f.b.	Dittus (1977)

[a]Sexual dimorphism from Harvey et al. (1986).
[b]From Appendix A–I, Smuts et al. (1986). M.m. f.b.=large multimale female-bonded; f. choice=female choice; large f. tran=large female transfer.

Table 5.3. *United States eating pattern groups by sex*

	Percentage of males	Percentage of females
I More dairy, soups – less sugary foods/beverages	14	14
II More nonsugary beverages – less dairy	16	23
III More eggs, legumes, nut cereals, grains	18	12
IV More meats, vegetables/fruits/juices, desserts	20	13
V More poultry – fewer red meats	12	16
VI More mixed protein, shellfish	12	12
VII More fish, fats, oils	8	10
Sample size (*N*)	7656	10306

Schwerin et al. *(1981)*

age of seven that it is shameful and bad luck to ask for additional fish with one's rice (Sharman *et al.* 1991). The Tswana have a highly elaborate set of rules about which parts of animals men and women may eat (Grivetti 1978). Specific meats are frequently forbidden to women on the basis of reproductive status (Bolton 1972; Ross 1978; White 1985). Tswana nursing mothers are fed fermented sorghum, milk, and greens (Grivetti 1978). Among the Donydji Australians (White 1985) nursing mothers are forbidden meat from water goannas, some birds, the echidna, some fish, and the black nosed file snake. Hurtado and Hill (1990) argue that hunting (in part), and the gathering of some dispersed but nutritious fruits, is denied women because of sexual harassment or assault by non-community males. They suggest that women do not engage in hunting because it endangers dependent offspring. This hypothesis is consistent with observations that females limit foraging activities to areas that are close to the home base (De Schlippe 1956; Brown 1970, 1976; White 1985; Hurtado and Hill 1990). Note, however, that Willoughby (1963) documented very diverse hunting among California Native American women, though in keeping with observations of White (1985) and Hurtado and Hill (1990), women's prey was typically smaller and found nearer to camp.

It is possible that sex differences in diet are a human universal. For example, Table 5.3 presents data from a factor analysis of the Ten-State and HANES surveys in the United States (Schwerin *et al.* 1981), in which distinct groups were identified by eating patterns. Women, it appears chose lower meat protein, higher calorie-density foods, and oddly, lower mineral diets than men. This pattern was recognized despite the strong living-together effect expected in spouses' diets (Kolonel and Lee 1981), and the lack of well-articulated gender differences in access to foods in our culture. A striking feature of the proximate analysis presentations of the HANES data is the relative and absolute low calcium and high vitamin A and C intake of women of childbearing age. Sex differences in diseases of the digestive system, such as gall bladder disease, which is two to five times more common in US females (Weiss *et al.* 1984), may suggest the presence of evolutionary repercussions of different diets between the sexes.

We think that the diet of women in the United States, being higher in calorie-density than the diet of men, may reflect the retention of human dimorphism based upon reproductive biology. The retention of dimorphic body composition, especially when measured in terms of body fat (Lancaster 1984), may also be an evolutionary legacy. If retention of body fat in females was critical to successful reproduction in our evolutionary past, selection for individuals seeking calorie-dense diets, as well as for females capable of successfully storing lipids, might well have been occurring. The persistence of this pattern in modern times, despite the relatively light reproductive costs because of our low birth rates and our limited breast-feeding practices, is striking. This is especially true in light of the absence of extensive nutritional deficiency in American culture, and the very limited labor investment needed in the quest for food.

Does gender inequality account for dietary differences in trace elements?

If sex differences in diet are primarily related to reproductive demands, the appeal to apply gender inequality models is compelling, but unfortunately premature. The relative contributions of gender inequality models, as well as models that stress reproductive demands, are limited because most workers make an *a priori* choice to use one model or the other. If we want to search for clear evidence of gender inequality in access to foodstuffs, the topic should be pursued in archaeological and ethnographic contexts. In pre-modern India, for example, ethnohistorical and contemporary ethnographic data consistently lead one to expect that gender inequality was a feature of the past. Indeed, the food prohibitions and sex-stereotyped behaviors we observe in present day societies may provide insight into our evolutionary past.

The recognized sex differences in food choice, common to virtually all the primates, can be argued to reflect the demands that lactation places on fat, calcium, sodium, and chloride reserves in females. For instance, we know that milk is particularly high in calcium, sodium, and chloride (Jelliffe and Jelliffe 1978). Where dairy products and access to salt are limited, pregnant and lactating women may have had difficulty maintaining necessary stores of these elements. Dietary preferences for plants with high mineral contents, for foliage or tubers over fruits, and for ashes, clays, and other direct sources of minerals, may have been an adaptive response under these conditions. Perceived differences in trace element levels between males and females in these populations would not be due to gender inequality, but rather, would have a biological basis. Geophagy is widely reported among modern humans, especially among pregnant women (Hochstein 1968; Farb and Armelagos 1980). An example is provided by Amazonian peoples who supplement a very low sodium diet with a variety of plant ash seasonings, which contributes significantly to their mineral nutrition (Coimbra 1985; Fleming-Moran *et al.* 1991). Greens (Fleuret 1979), clays, and bones cooked in acid sauces (Beiser *et al.* 1974; Farb and Armelagos 1980), and even cremated human bone (Chagnon 1974), may be important sources of calcium, iron and other minerals for groups that do not use milk products.

Studies of trace element differences in ancient bones that allude to the presence of gender inequality in the past are largely products of our modern interest in gender issues. We were told that 'real men don't eat quiche' (Feirstein 1982). While this may well be true, at least if the quiche is spinach, the reasons may lie less in gender inequality than in reproductive biology. The wealth of recent literature specifically critiquing 'deterministic' views of sex differences (e.g., Bacus *et al.* 1993; Schiebinger 1993) has stimulated a reevaluation of the meaning of 'difference'. We should not lose sight, however, of fundamental and undeniable biological differences between males and females as we explore gender issues in the past. The alternative biological science that Schiebinger (1993) imagines – a biology not based on reproductive differences as its organizing principle – would be a poor substitute for our present explanatory models.

References

Adair LS (1987) Nutrition in the reproductive years. In FE Johnston (ed.), *Nutritional Anthropology*. New York: Liss, pp.119–54.
Ambrose S (1993) Isotopic analysis of paleodiets: methodological and interpretive considerations. In MK Sandford (ed.), *Investigations of Ancient Human Tissue: Chemical Analyses in Anthropology*. Langhorne: Gordon and Breach.
Bacus EA, Barker AW, Bonevich JD, Dunavan SL, Fitzhugh JB, Gold DL, Goldman-Finn NS, Griffin W, and Mudar KM (Eds.) (1993) *A Gendered Past: A Critical Bibliography of Gender in Archaeology*. Michigan: University of Michigan Museum of Anthropology Technical Report No.25.
Badrian A and Badrian N (1984) Social organization of *Pan paniscus* in the Lomako Forest, Zaire. In RL Susman (ed.), *The Pygmy Chimpanzee: Evolutionary Biology and Behavior*. New York: Plenum Press, pp.325–46.
Barton RA, Whiten A, Burne RW, and English M (1993) Chemical composition of baboon plant foods: implications for the interpretation of intra- and interspecific differences in diet. *Folia Primatologica* **61**:1–20.

Beck LA (1985) Bivariate analysis of trace elements in bone. *Journal of Human Evolution* 14:493–502.

Beiser M, Burr WA, Collomb H, and Ravel JL (1974) Pobough Lang in Senegal. *Social Psychiatry* 9:123–9.

Blakely RL (1989) Bone strontium in pregnant and lactating females from archeological samples. *American Journal of Physical Anthropology* 80:173–88.

Bolton JM (1972) Food taboos among the Orang Asli in West Malasia: a potential nutritional hazard. *American Journal of Clinical Nutrition* 25:789–99.

Brown JK (1970) A note on division of labor by sex. *American Anthropologist* 72:1073–8.

Brown JK (1976) An anthropological perspective on sex roles and subsistence. In MS Teitelbaum (ed.), *Sex Differences*. Garden City: Anchor Books, pp.122–37.

Buikstra JE (1984) Lower Illinois River region: a prehistoric context for the study of ancient diet and health. In MN Cohen and GJ Armelagos (eds.), *Paleopathology at the Origins of Agriculture*. Orlando: Academic Press, pp.215–34.

Buikstra JE, Bullington JE, Charles DK, Cook DC, Frankenberg SR, Konigsberg LW, Lambert JB, and Xue L (1987) Diet, demography, and the development of horticulture. In WF Keegan (ed.), *Emergent Horticultural Economies in the Eastern Woodlands*. Carbondale: Southern Illinois University Press, pp. 67–85.

Buikstra JE, Frankenberg S, Lambert JB and Xue L (1989) Multiple elements: multiple expectations. In TD Price (ed.), *The Chemistry of Prehistoric Human Bone*. Cambridge: Cambridge University Press, pp.155–210.

Burton JH and Price TD (1990) Ratio of barium to strontium as a paleodietary indicator of consumption of marine sources. *Journal of Archaeological Science* 17:547–57.

Burton JH and Wright LE (1995) Nonlinearity in the relationship between bone Sr/Ca and diet: paleodietary implications. *American Journal of Physical Anthropology* 96: 273–82.

Cartmill M (1993) *A View to a Death in the Morning: Hunting and Nature Through History*. Cambridge, Mass.: Harvard University Press.

Chagnon NA (1974) *Studying the Yanomamo*. New York: Holt, Rinehart and Winston.

Clutton-Brock TH (1974) Activity patterns of red colobus (*Colobus badiustephrosceles*). *Folia Primatologica* 21:161–87.

Clutton-Brock TH (1977) Some aspects of intraspecific variation in feeding and ranging behaviour in primates. In TH Clutton-Brock (ed.), *Primate Ecology: Studies of Feeding and Ranging Behaviours in Lemurs, Monkeys, and Apes*. London: Academic Press, pp. 539–66.

Coimbra CEA (1985) Estudos de ecologia humana entre os Suruı do Parque Indigena Aripuana, Rondonia. *Boletin do Museu Paraense Emilio Goeldi* 2:9–87.

Cook DC (1979) Subsistence base and health in prehistoric Illinois Valley: evidence from the human skeleton. *Medical Anthropology* 3:109–24.

De Schlippe P (1956) *Shifting Cultivation in Africa: The Zande System of Agriculture*. London: Routledge and Kegan Paul.

Dittus W (1977) The social regulation of population density and age-sex distribution in the toque monkey. *Behaviour* 63:281–322.

Farb P and Armelagos GJ (1980) *Consuming Passions: The Anthropology of Eating*. Boston: Houghton Mifflin.

Feirstein B (1982) *Real Men Don't Eat Quiche*. New York: Pocket Books.

Fleming-Moran M, Santos RV, and Coimbra CEA (1991) Blood-pressure levels of the Surui and Zoro Indians of the Brazilian Amazon. *Human Biology* **63**:835–61.

Fleuret A (1979) Methods for the evaluation of the role of fruits and wild greens in Shambaa diet: a case study. *Medical Anthropology* 3:249–69.

Fossey D and Harcourt AH (1977) Feeding ecology of free-ranging mountain gorillas (*Gorilla gorilla beringei*). In TH Clutton-Brock (ed.), *Primate Ecology: Studies of Feeding and Ranging Behaviours in Lemurs, Monkeys, and Apes*. London: Academic Press, pp. 415–47.

Galdikas BMF and Teleki G (1981) Variations in subsistence activities of female and male pongids: new perspectives on the origins of hominid labor division. *Current Anthropology* **22**: 241–56.

Gautier-Hion A (1980) Seasonal variations of diet related to species and sex in a community of *Cercopithecus* monkeys. *Journal of Animal Ecology* **49**:237–69.

Goodall J (1986) The Chimpanzees of Gombe: Patterns of Behavior. Cambridge, Mass.: Harvard University Press.

Grivetti LE (1978) Nutritional success in a semi-arid land: examination of Tswana agro-pastoralists of the eastern Kalahari, Botswana. *American Journal of Clinical Nutrition* **31**:1204–12.

Harding RSO (1973) Predation by a troop of olive baboons (*Papio anubis*). *American Journal of Physical Anthropology* **38**:587–91.

Harding RSO and Strum SC (1976) The predatory baboons of Kekopey. *Natural History* **85**:46–53.

Harrison MJS (1983) Age and sex differences in the diet and feeding strategies of the green monkey, *Cercopithecus sabaeus*. *Animal Behavior* **31**:969–77.

Harvey PH, Martin RD, and Clutton-Brock TH (1986) Life histories in comparative perspective. In BB Smuts, DL Cheney, RM Seyfarth, RW Wrangham, and TT Struhsaker (eds.), *Primate Societies*. Chicago: University of Chicago Press, pp. 181–96.

Hochstein G (1968) Pica: a study in medical and anthropological explanation. In T Weaver (ed.), *Essays on Medical Anthropology*. Southern Anthropological Society Proceedings, No.1. Athens: University of Georgia Press, pp. 88–96.

Hurtado AM and Hill KR (1990) Seasonality in a foraging society: variation in diet, work effort, fertility, and sexual division of labor among the Hiwi of Venezuela. *Journal of Anthropological Research* **46**:293–347.

Jelliffe DB and Jelliffe EFP (1978) *Human Milk in the Modern World: Psychosocial, Nutritional, and Economic Significance.* New York: Oxford University Press.

Katz SH, Hediger ML, and Valleroy LA (1974) Traditional maize processing techniques in the New World. *Science* **184**:765–73.

Katzenberg MA (1984) *Chemical Analysis of Prehistoric Human Bone from Five Temporally Distinct Populations in Southern Ontario.* National Museum of Man – Mercury Series No.129. Ottawa: National Museums of Canada.

Klepinger LL (1984) Nutritional assessment from bone. *Annual Review of Anthropology* **13**:75–96.

Kolonel LN and J Lee (1981) Husband–wife correspondence in smoking, drinking, and dietary habits. *American Journal of Clinical Nutrition* **34**:99–104.

Lambert JB, Vlasak SM, Thometz AC, and Buikstra JE (1982) A comparative study of the chemical analysis of ribs and femurs in Woodland populations. *American Journal of Physical Anthropology* **59**:289–94.

Lancaster JB (1984) Evolutionary perspectives on sex differences in the higher primates. In AS Rossi (ed.), *Gender and the Life Course*. Chicago: Aldine, pp. 3–27.

Lovell NC, Nelson DE, and Schwarcz HP (1986) Carbon isotopes in paleodiet: lack of age or sex effect. *Archeometry* **28**:51–5.

Mahaney WC, Milner MW, Sanmugadas K, Hancock RGV, Aufreiter S, Wrangham RW and Pier H (1997) Analysis of geophagy soils in Kibale Forest, Uganda. *Primates* **38**(2):159–76.

Norton GW (1987) Baboon diet: a five-year study of stability and variability in the plant feeding and habitat of the yellow baboons (*Papio cynocephalus*) of Mikumi National Park, Tanzania. *Folia Primatologica* 48:78–120.

Pollock JI (1977) The ecology and sociology of feeding in *Indri indri*. In TH Clutton-Brock (ed.), *Primate Ecology: Studies of Feeding and Ranging Behaviours in Lemurs, Monkeys, and Apes.* London: Academic Press, pp. 38–69.

Pollock, JI (1979) Female dominance in Indri indri. *Folia Primatologica* 31:143–64.

Price TD (1989) *The Chemistry of Prehistoric Human Bone.* Cambridge: Cambridge University Press.

Price TD, Swick RW, and Chase EP (1986) Bone chemistry and prehistoric diet: strontium studies of laboratory rats. *American Journal of Physical Anthropology* **70**:365–75.

Radosevich SC (1993) The six deadly sins of trace element analysis: a case of wishful thinking in science. In MK Sandford (ed.), *Investigations of Ancient Human Tissue: Chemical Analyses in Anthropology.* Langhorne: Gordon and Breach, pp. 269–332.

Reidhead VA (1981) *A Linear-Programming Model of Prehistoric Subsistence Organization: A Southeastern Indiana Example.* Indianapolis: Indiana Historical Society.

Remis M (1995) Effects of body size and social context on the arboreal activities of lowland gorillas in the Central African Republic. *American Journal of Physical Anthropology* **97**:413–33.

Richard A (1986) Malagasy prosimians: female dominance. In BB Smuts, DL Cheney, RM Seyfarth, RW Wrangham, and TT Struhsaker (eds.), *Primate Societies.* Chicago: University of Chicago Press, pp. 25–33.

Rodman PS (1977) Feeding behavior of orang-utans of the Kutai Nature Reserve, East Kalimantan. In TH Clutton-Brock (ed.), *Primate Ecology: Studies of Feeding and Ranging Behaviours in Lemurs, Monkeys, and Apes.* London: Academic Press, pp. 383–413.

Rose LM (1994) Sex differences in diet and foraging behavior in white-faced capuchins (*Cebus capucinus*). *International Journal of Primatology* **15**: 95–114.

Rosenberg EM (1980) Demographic effects of sex-differential nutrition. In N Jerome, NW Jerome, RF Kandel, and GH Pelto (eds.), *Nutritional Anthropology.* Redgrave: Pleasantville.

Ross EB (1978) Food taboos, diet, and hunting strategy: the adaptation to animals in Amazon cultural ecology. *Current Anthropology* **19**:1–36.

Runia LT (1987) *The Chemical Analysis of Prehistoric Bones: A Paleodietary and Ethnoarchaeological Study of Bronze Age West-Friesland.* Oxford: British Archaeological Reports International Series No. 363.

Sanford MK (1993) *Investigations of Ancient Human Tissue.* Philadelphia: Gordon and Breach.

Sauther ML and Nash LT (1987) Effect of reproductive state and body size on food consumption in captive *Galago senegalensis braccatus. American Journal of Physical Anthropology* **73**: 81–8.

Schoeninger MJ and Peebles CS (1981) Effect of mollusc eating on human bone strontium levels. *Journal of Archaeological Science* **8**:391–7.

Schiebinger L (1993) *Nature's Body: Gender in the Making of Modern Science.* Boston: Beacon.

Schwerin HS, Stanton JL, Riley AM, Schaefer AE, Leveille GA, Elliot JG, Warwick KM, and Brett BE (1981) Food eating patterns and health: a reexamination of the Ten-State and HANES I surveys. *American Journal of Clinical Nutrition* **34**:568–80.

Sharman A, Theophano J, Curtis K, and Messer E (1991) Introduction. In A Sharman (ed.), *Diet and Domestic Life in Society.* Philadelphia: Temple University Press, pp. 3–14.

Sillen A and Kavanagh M (1982) Strontium and paleodietary research: a review. *Yearbook of Physical Anthropology* **25**:67–90.

Smuts BB, Cheney DL, Seyfarth RM, Wrangham RW, and Struhsaker TT (Eds.) (1986) *Primate Societies.* Chicago: University of Chicago Press.

Speth JD (1989) Early hominid hunting and scavenging: the role of meat as an energy source. *Journal of Anthropological Archaeology* **2**:1–31.

Strum SC (1975) Primate predation: interim report on the development of a tradition in a troop of olive baboons. *Science* **187**:755–7.

Ubelaker DH, Katzenberg MA, and Doyon LG (1995) Status and diet in precontact highland Ecuador. *American Journal of Physical Anthropology* **97**:403–12.

van Lawick-Goodall J (1968) The behaviour of free-living chimpanzees in the Gombe Stream Reserve. *Animal Behaviour Monographs* **1**(3):165–311.

Waser P (1977) Feeding, ranging, and group size in the mangabey *Cercocebus albigena.* In TH Clutton-Brock (ed.), *Primate Ecology: Studies of Feeding and Ranging Behaviours in Lemurs, Monkeys, and Apes.* London: Academic Press, pp. 183–222.

Watts D (1985) Relations between group size and composition and feeding competition in mountain gorilla groups. *Animal Behavior* **33**:72–85.

Weiss KM, Ferrel RE, and Harris CL (1984) A New World syndrome of metabolic diseases with a genetic and evolutionary basis. *Yearbook of Physical Anthropology* **27**:153–8.

White CD and Schwarcz HP (1989) Ancient Maya diet: as inferred from isotopic and elemental analysis of human bone. *Journal of Archaeological Science* **16**:451–74.

White N (1985) Sex differences in Australian aboriginal subsistence: possible implications for the biology of hunter-gatherers. In J Ghesquire, RD Martin, and F Newcombe (eds.), *Human Sexual Dimorphism.* Philadelphia: Taylor and Francis, pp. 323–61.

Willoughby N (1963) Division of labor among Indians of California. *University of California Archaeological Survey Reports* **60**:7–79.

Wing E and AB Brown (1979) *Paleonutrition: Method and Theory in Prehistoric Foodways.* New York: Academic Press.

Wrangham RW (1977) Feeding behaviors of chimpanzees in Gombe National Park, Tanzania. In TH Clutton-Brock (ed.), *Primate Ecology. Studies of Feeding and Ranging Behaviours in Lemurs, Monkeys and Apes.* London: Academic Press, pp. 503–38.

Wrangham RW, McGrew WC, de Waal FBM, Heltne PG, and Marquardt LA (1994) *Chimpanzee Cultures.* Cambridge, Mass.: Harvard University Press.

Wright P (1984) Biparental care in *Aotus trivigatus* and *Callicebus moloch.* In M Small (ed.), *Female Primates: Studies by Women Primatologists.* New York: Alan R. Liss, pp. 59–75.

6

Male–female immune reactivity and its implications for interpreting evidence in human skeletal paleopathology

DONALD J. ORTNER

Infectious disease undoubtedly has played a significant role in the evolution of modern humans. Archaeological human remains offer insight into some aspects of this process, but interpreting the evidence is challenging since some of the potential factors affecting the process are unknown and those that are known are difficult or impossible to control. The purpose of this chapter is to explore a few of these factors and to offer a theoretical model that may stimulate testable hypotheses for future research.

As paleopathologists have developed an interpretative understanding of disease manifested in skeletal remains, attempts have been made to reconstruct the dynamic relationships between infectious agents and human populations as they vary in both time and space. For example, there are reports about the relative health of skeletal samples based on the prevalence of disease conditions in skeletal material (e.g., Goodman *et al.* 1984; Powell 1988; Rose and Hartnady 1991). These reports indicate the robusticity of recent paleopathological research and certainly point in an important direction for research. However, there is a danger that interpretation of skeletal data will proceed more rapidly than is warranted by the development of sound theory. Wood *et al.* (1992) have explored some models for inferring health in archaeological human populations that articulate the need for caution.

Skeletal participation in infectious disease occupies an intermediate position on the continuum of immune responses that ranges from rapid death at one end to complete recovery with no adverse complications or lingering effects at the other. The important, although somewhat paradoxical, point is that the human body needs a relatively good immune response to have bone involvement from infection, but not too good so as to allow recovery with no skeletal involvement.

Many factors affect an individual's potential to make the transition from exposure to an infectious agent to morbidity and, possibly, death (Table 6.1). Clearly, the necessary factor is the exposure to the agent which will be greatly affected by human culture. For example, in situations where men do most of

Table 6.1. *Factors that affect the expression of infectious disease*

• Age of onset	• Exposure to infectious agents
• Nutritional status of the patient	• Biology of the infectious agent
• Immune response of the patient	• Size of the inoculum
• General health of the patient	• Social conditions
• Age of the patient	• Environmental conditions
• Point of entry	• Efficacy of treatment

Table 6.2. *Modern male–female prevalence ratios for various infectious diseases*

Disease	Ratio	Source
Syphilis	3:1	Pusey (1933)
Bejel	M < F	Hoeprich (1989)
	1:1	Hudson (1946)
Yaws	1:1	Toure (1985)
Osteomyelitis, hematogenous	3:1	Trueta (1959)
Tuberculosis	2:1	International Union Against Tuberculosis (1964)
Leprosy	2–3:1	Meyers (1992)
Brucellosis	5:1	Hall and Khan (1989)
Leishmaniasis	M > F	Weigle *et al.* (1993)
Actinomycosis	2:1	Guidry (1971)
Blastomycosis	6–10:1	Utz (1989)

the work in planting crops, exposure to infectious agents in the soil will be greater for men than women. Fungus (mycotic) infections, such as blasto-mycosis, are much more common in men than women partly because of increased male exposure to fungi during farming activities (Table 6.2).

Only a small percentage of individuals who have an infectious disease will exhibit evidence of the disease in the gross anatomy of the skeleton. This means that in a skeletal sample, those skeletons showing evidence of infection may not be representative of all individuals in the total sample who had infectious disease. Virtually all the infectious diseases encountered in human skeletal remains are the result of chronic conditions in which the patient survived with the disease for many years and skeletal involvement was late in the disease process.

Female immune reactivity and maternal mortality

The specialized nature of skeletal disease is further complicated by the fact that there are sex differences in the immune response to disease. The immune response of women to infectious disease is greater and more effective than the response of men (Ahmed *et al.* 1985; Grossman 1985; Talal 1992; Roitt 1994). Two major factors for this difference are: (1) selective pressure associated with the hazards of pregnancy and childbirth in women; and (2) the sex-related differences in physiology, particularly in sex hormones. The enhanced immune response in women has the potential of differentially affecting the prevalence of infection amongst the sexes in archaeological samples.

The biological role of women as child bearers, involves risks for both morbidity and mortality. Factors contributing to mortality associated with pregnancy and childbirth include both direct and indirect threats to life. Today direct factors include infection, hemorrhage, eclampsia, obstructed labor, ectopic pregnancy, complications arising from attempted abortion, malnutrition, and complications arising from treatments during pregnancy, childbirth and the period immediately following birth. Currently, in developing countries direct factors account for 50–98% of maternal mortality. Of these factors, infection and hemorrhage cause about 50% of the deaths (*WHO Chronicle* 1986). Indirect causes of maternal mortality include conditions present at the time of conception that are exacerbated by pregnancy. The most common of these are infectious diseases, hemopoietic diseases (e.g., anemia), heart disease, diabetes, and hypertension.

Pregnancy is a significant cause of death in developing countries. Royston and Lopez (1987) report that mortality per pregnancy ranges between 0.1% in Mauritius and 2.0% in Ethiopia. In developing countries, a woman's probability of dying from pregnancy ranges from about 1:15 to 1:70 (World Health Organization 1987). Between 15 and 49 years of age, maternal mortality accounts for 28% of female deaths in Karachi, Pakistan (Fortney *et al.* 1987).

It is likely that maternal mortality and adaptive responses to it have been a significant component of human evolution. The epochal changes in human society during the last 10,000 years, including domestication of plants and animals, increasing sedentism, and the development of cities, have created conditions that have made pregnancy an even greater risk by increasing exposure to infectious agents. Enhanced female immune reactivity may be one of the adaptive mechanisms in response to increased exposure to the hazards of pregnancy, particularly infection. Although sex differences in immune reactivity are seen in other mammals, increased selective pressure may have resulted in additional adaptive change in human groups during the Holocene.

Gender as a factor in malnutrition and morbidity

Another factor that might affect sex-related prevalence of infection is cultural differences in access to food. In developing countries, men may have preferential access to both the quantity and quality of food (Carloni 1981; Gittelsohn 1991). Boys are next in access with women and girls last. There are exceptions to this pattern, particularly in sub-Saharan Africa (Pennington and Harpending 1993:58–9). Differential access to food becomes even more severe during famine, putting females at a much greater disadvantage (Rosenberg 1980:188–9; Hurtado and Hill 1990:320). It seems plausible that sex preference in food access has a long history.

The link between malnutrition and increased vulnerability to infectious disease is well established. Indeed malnutrition and infection exacerbate the vulnerability to each other (Chen *et al.* 1981:59). However, despite greater evidence of malnutrition among female children in the Bangladesh study (Chen *et al.* 1981), there was no significant difference in infection rates. This suggests similar exposure of males and females to infectious agents (Chen *et al.* 1981:64). In the same study, male children are brought to the free clinic more often than female children. While this is interpreted as evidence of preferential treatment of male children, it may indicate that female children less often require modern medical intervention to recover from infectious disease (Chen *et al.* 1981:66). This suggests that differential access to food does not affect differential infectious morbidity by sex, but may affect a child's ability and speed to recover. An extension of this argument is that malnutrition may have a greater effect on mortality rates than on morbidity in infectious disease.

Famine is one of the complicating factors in interpreting evidence of disease in skeletal samples. It is probable that famine and resulting malnutrition will increase mortality as a complication of infection. Morbidity will also be affected. Since the body's immune response is depressed in malnutrition, the likelihood of dying before a disease becomes chronic is increased. Diseases in an individual with adequate nutrition will more likely become chronic and affect the skeleton. In theory, therefore, periods of malnutrition can lead to decreased evidence of skeletal disease in archaeological populations, creating the impression of better health when, in fact, individuals were very sick and died quickly.

Hypothetical models of male and female immune reactivity

Theoretically, if all factors were equal, women might be expected to survive chronic stages of infectious disease more often than men. One of the conceptual problems needing resolution, however, is the optimal location and range on the continuum of the immune response that will result in skeletal lesions. Understanding this dimension of skeletal pathology is important in interpreting sex differences in the prevalence of skeletal disease. It is also critical in making inferences about the relative health of a human population based on skeletal evidence of disease.

To assist in understanding the effect of differential male–female immune reactivity, a few assumptions need to be made as we explore some of the variables involved in producing skeletal manifestations in response to infectious disease. In Figure 6.1, a graph is presented in which the X axis is some measure of immune response ranging from 'poor' to 'good'. The Y axis is a measure of the number of individuals at any point on the X axis. Given the known sex difference in immune reactivity, the male and female subsamples of the population would probably be arranged as two partially superimposed normal distributions. The mean ($\bar{x}_1$) of the male subsample would be positioned more towards the 'poor' end of the scale than the female mean ($\bar{x}_2$). Among the unknown variables are the location of the two means on the immune response scale and the distance between $\bar{x}_1$ and $\bar{x}_2$.

An additional variable is the hypothetical range on the immune response scale where skeletal involvement occurs. At the 'poor' end of the scale we may designate R_1 as the point below which skeletal involvement does not occur and death is a typical event. R_2 is the point on the other end of the

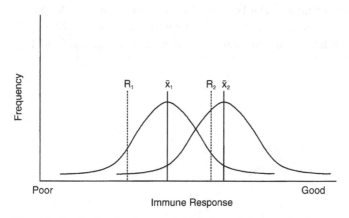

Fig. 6.1 Graph showing hypothetical male and female distributions on an immune response scale from 'Poor' to 'Good'. $\bar{x}_1$ designates the mean of the male and $\bar{x}_2$ the mean of the female distribution. R_1 is a theoretical point below which skeletal manifestations of disease do not occur and death is a typical event. R_2 is the point where no skeletal disease occurs because of complete recovery from infectious disease. R_1 and R_2 are positioned to create a male:female ratio of approximately 3:1 similar to clinical ratios of infectious disease.

scale beyond which skeletal manifestations do not occur because of complete recovery from infectious disease. While the argument that such lines probably exist is plausible, where on the scale R_1 and R_2 occur is not known. Furthermore, we do not know the range between R_1 and R_2. It is also theoretically possible that the location and range for variables $\bar{x}_1$ and $\bar{x}_2$, as well as R_1 and R_2, on the immune response scale vary in different infectious diseases and between human populations.

If, for the moment, we assume that all other variables (most particularly, exposure to infectious agents) that affect the expression of infectious disease in the human skeleton are constant and that the distribution approximates normality, it is apparent that the position and range of $\bar{x}_1$ and $\bar{x}_2$, as well as R_1 and R_2, on the scale will affect the sex ratios for the prevalence of infectious disease in a skeletal sample. If the positions and ranges of R_1 and R_2 are more toward the 'poor' end of the immune response scale (Figure 6.1), proportionately more males would be included in the range most likely to result in skeletal manifestations of infection. A shift in the positions and ranges of R_1 and R_2 toward the 'good' end of the scale (Figure 6.2) would increase the proportionate number of females that express skeletal disease.

In the paragraphs that follow, I will use published data on sex ratios for infectious diseases in modern clinical reports and statistics on periostitis in archaeological samples (which is most commonly caused by infection), to explore options for interpreting data on infection in archaeological human populations. The intent is to clarify some of the complexity in the interpretive process and to caution against overly simplistic interpretations. Another objective is to show how archaeological populations can aid in understanding

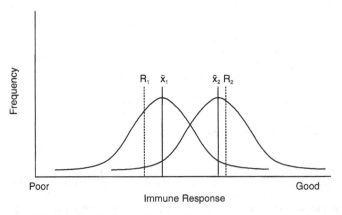

Fig. 6.2 Graph with R_1 and R_2 positioned to create a ratio of approximately 1.5:1, similar to male:female ratios of infectious disease in an archaeological sample.

Table 6.3. *Prevalence of periostitis in archeological human skeletal samples*

Locality	Male cases of periostitis relative to males in the sample (%)	Female cases of periostitis relative to females in the sample (%)	Source
Northgate, Kent, UK (AD 1084–1560)	21/261 (8)	11/248 (4)	Anderson (1995, unpublished data)
Moundville, Alabama, USA (AD 1059–1550)	13/192 (7)	15/242 (6)	Powell (1988)
Plains, USA (Woodland and Historic)	13/299 (4)	7/222 (3)	Repatriation Office National Museum of Natural History (1997, unpublished data)
Georgia and Florida, USA Precontact preagricultural	3/32 (9)	2/47 (4)	Larsen (see Chapter 11)
Georgia and Florida, USA Precontact agricultural	22/93 (23)	32/133 (24)	Larsen (see Chapter 11)
Georgia and Florida, USA Early contact	3/13 (23)	1/7 (14)	Larsen (see Chapter 11)
Georgia and Florida, USA Late contact	25/36 (70)	23/35 (65)	Larsen (see Chapter 11)

evolutionary dynamics that have led to differences we see today in human disease responses.

Male–female prevalence ratios in infectious disease

The data in Table 6.2 are from clinical reports on the prevalence ratios between males and females for a variety of infectious diseases. It is apparent in the ratios for bejel that different reports provide different ratios. Ratios are likely to vary with time, geography, culture, and changes in diagnostic criteria. However, they do tend to cluster and consistently show greater male morbidity for infection. The data in Table 6.3 provide the number of cases of periostitis relative to the number of males and females in the skeletal samples.

The data in the two tables are from different types of samples. Table 6.2 is based on the prevalence of various infectious diseases in a clinical setting. Most of the patients did not have nor were ever likely to acquire skeletal manifestations of the various infectious diseases. In Table 6.3 all cases of infection are identified as pathological because of skeletal evidence of disease. Table 6.3 undoubtedly represents a narrower range in the disease spectrum, with an emphasis on individuals with chronic infectious disease.

In addition to the fundamentally different nature of the samples used for Tables 6.2 and 6.3, exposure to infectious agents will vary between societies and between males and females depending on their culturally defined roles within the society. Both tables indicate that males have higher morbidity than females. One question is whether this is simply the result of greater exposure to infectious agents among males or a more effective immune response among females. In some infectious diseases, such as mycotic infection, a case can be made for greater exposure to infectious agents by males in agricultural societies. However, in at least some cultural contexts and in some age ranges, males and females seem to have equal exposure to infectious agents, but males seem more vulnerable to disease (Chen *et al.* 1981). Most likely, exposure of males and females varies between infectious agents, age categories, and cultural contexts, but overall is probably close to equal. Given this, I argue that the male predominance in morbidity in both clinical and archaeological contexts is at least partially due to enhanced female immune reactivity.

The other factors affecting vulnerability to infectious disease in males and females are less clear. For example, nutritional differences resulting from different access to food by male and female children seem to have no observable difference in the prevalence of disease in some social contexts (Chen *et al.* 1981). Our ability to control for any of these variables in an archeological sample is limited to say the least. Nevertheless, a useful starting point is to compare sex ratios for infectious disease in archeological samples with the ratios reported in the clinical literature.

Unfortunately, data on infection in archeological samples is inadequate to provide the sample sizes within infectious disease categories that permit specific comparisons with data from a clinical context. Table 6.3 provides data on infection, detected by the presence of periosteal reactions (see Figure 10.2), or periostitis, in archeological samples. These general data can provide insight and direction for additional research, but should not be considered definitive. Much larger samples and more specific diagnoses are needed to provide a better basis for comparison with modern clinical data.

Implications for interpreting evidence of disease in archaeological human skeletons

The data in Table 6.3 indicate that in three different archaeological human skeletal samples males are more often affected by lesions probably attributable to infection than females. The ratios range from slightly more than 1:1 to 2:1. We do not know which infectious diseases caused these conditions, and this makes direct comparison with the ratios in Table 6.2 problematic. Nevertheless, it does seem that the range of ratios in Table 6.3 is at the low end of that expected on the basis of a direct extrapolation from Table 6.2. In Table 6.2, both the nonvenereal forms of treponematosis (bejel and yaws) have a low ratio of 1:1, but are the only chronic diseases in which this is the case. For all other diseases male prevalence is greater than female and in some cases, such as in blastomycosis, it is considerably greater.

There are several factors that could contribute to the lower ratios found in the archaeological samples. One of these may be that more of the infectious conditions affecting the archaeological populations have a low male-female ratio. Another possibility is that clinically identified infectious conditions are likely to represent a different range on the immune response scale, and thus may include a different proportion of males and females. In this situation, the lower ratios of male–female disease prevalence could be caused by skeletal diseases whose R_1 and R_2 lie toward the 'good' end of the immune response scale (Figure 6.2), and thus would include a greater number of females in the sample of individuals having infectious disease. A third possibility is that the lower ratios apparent in the archaeological skeletal samples are due to a higher proportion of females reaching the chronic stages of infectious disease because of their enhanced immune response.

Implications for research in paleopathology

Gender differences in immune reactivity are important factors in interpreting the paleoepidemiology of infectious disease. They are also important in our understanding of evolutionary mechanisms that have developed with changes in infectious agents and the social context of human groups. In both mice and hamsters, the female has greater immune reactivity than the male (Grossman 1985:257; Talal 1992). In a survey on infant mortality in captive primate colonies (Bustamente and Ortner 1995, unpublished data), infant mortality was higher in male chimpanzees, gorillas, orangutans, and gibbons. It is possible that this is because of greater immune reactivity in females. The fact that enhanced female immunity is not limited to the human species suggests that it is a fundamental adaptive strategy in many mammals. The evolutionary mechanisms operating to produce male–female immunological differences might be due to particular selective pressures associated with childbearing. During pregnancy the female immune system is depressed to minimize the possibility of rejecting the developing fetus (Grossman 1985:259). Women are particularly vulnerable to infection during pregnancy and immediately following birth. This makes infection a significant factor in maternal morbidity and mortality in areas and time periods where hygiene is poor and access to effective therapy for infection is unavailable.

In this context, one explanation for enhanced female immunity is an evolutionary balancing mechanism for the increased risk associated with pregnancy and childbirth. Since a significant number of women will die as a result of pregnancy, women need an enhanced response to infection when they are not pregnant so that more will survive the other challenges to life and ensure sufficient numbers of females to maintain the species.

An intriguing question arises – how much immune reactivity is too much? Disease can occur as a result of too little or too much reaction by the immune system. One dimension of the pathogenesis of miliary tuberculosis is an overly aggressive attack by the cells of the immune system on the tubercle bacillus in the lung. The excessive destruction of lung tissue in this process contributes to high mortality in this disease. Another negative expression of enhanced immune reactivity is autoimmune diseases that affect women more commonly than men. The greater prevalence of age-linked autoimmune diseases among women today is partly due to the increase in life expectancy seen over the past two centuries. Enhanced immune reactivity in women increases the potential for failure of the immune system's control mechanisms, resulting in the body's inability to recognize its own tissues and the body's consequent attempts to destroy them.

The example of autoimmune disease in women provides a partial explanation of why the immune reactivity of males does not match that of females. We see in the evolution of the female immune system that enhanced immune reactivity, as well as suppressed reactivity, to infectious agents carries risks. The human male balancing strategy does not involve the great risks associated with maternity. This may have provided a reduction in evolutionary pressures for enhancing the male immune response. Furthermore, male sexual function is related to increased levels of gonadal steroids, which tend to depress immune reactivity, while female sex hormones enhance immune reactivity (Grossman 1985).

Conclusions

The difference in male–female ratios between the archaeological samples and the clinical samples is unlikely to be attributable entirely to the increased prevalence of infectious diseases that have low ratios. The implication of this, as well as the fact of variability in sex ratios between various infectious diseases, is that individuals with and without skeletal disease represent different populations.

Infectious disease that affects the skeleton is relatively uncommon. Among tuberculosis patients less than 5% will have any skeletal involvement (Ortner and Putschar 1981:142). In leprosy, skeletal pathology affects between 3–5% of patients (Resnick and Niwayama 1988:2618). In individuals with treponematosis, estimates of skeletal involvement range from between 1% and 5% (Steinbock 1976: 143; Ortner and Putschar 1981:180–2). Higher figures (15%) have been reported for yaws (Hoeprich 1989:1025), but bone involvement is still uncommon. These percentages depend heavily on data from conventional radiology and may not reflect the more subtle expressions of skeletal involvement. However, individuals with infectious skeletal disease are likely to represent a different range on the immune response scale than populations of individuals with general infectious disease. The different immune response of males and females may provide an additional variable contributing to the contrast between skeletal and clinical samples.

One of the questions in pathogenesis raised earlier in this chapter is the optimal range and location on the immune response scale for the expression of skeletal pathology in infectious disease. Given the greater prevalence of male skeletons exhibiting infectious disease, it seems likely that the location on the immune response scale for the expression of skeletal infectious disease is toward the 'poor' end of the scale, thus including more of the male distribution. However, the current inability to control for other variables, particularly sex differences in exposure, calls for further thought and research.

Admittedly the models explored in the preceding paragraphs need development and refinement. What I have tried to demonstrate is that individuals with skeletal disease in either a clinical or archaeological sample have different immune responses to infection than individuals without skeletal disease, and that the more effective immune response to infection by females further complicates this difference. The implications of this observation, relative to interpretation of data on infectious disease in archeological populations, needs careful evaluation. Some research on disease in archaeological samples makes credible arguments for population changes in health based on the prevalence of infectious disease in skeletal samples. Despite the plausibility and quality of this research, I remain concerned that there may be systematic bias in interpreting evidence of infection in archaeological populations. The reason for this is that individuals with skeletal disease represent a very limited range on the immune response scale, and this range is very different from the range represented in living populations responding to infectious agents. It may be possible to reconstruct the location and range of individuals with skeletal disease on the basis of both clinical and archaeological data, along with careful refinement of appropriate models.

An important phase in the development of any scientific discipline is to thoroughly understand the various ways that data can be interpreted. In this chapter I have sought to explore ways to interpret evidence of infectious disease in archaeological populations, and I have elucidated some of the problems. I have also highlighted the fact that data on infectious disease in archaeological populations reveal information about the pathogenesis and the evolution of human immune reactivity. These insights are provided by skeletal responses to chronic infections and the sex differences that are associated with them.

Acknowledgments

The author is grateful for the advice of Dr. Lee-Ann Hayek, statistician, National Museum of Natural History, Smithsonian Institution, in the interpretation of data in Tables 6.2 and 6.3, and for her helpful input regarding the mathematical modeling explored in the report. Mr. David Paul Kuwayama, who had recently completed his undergraduate degree at the University of Wisconsin, was a volunteer research assistant during the preparation of this paper. He provided excellent assistance with the review of the biomedical literature and provided challenging questions and stimulating discussion of many of the ideas explored in the preceding paragraphs. Drs. George R. Milner, James W. Wood, and Kenneth M. Weiss, Pennsylvania State University, read a draft of the manuscript and offered suggestions that improved the content in situations where the author was able to incorporate them. Esther M. Sternberg, Chief, Section on Neuroendocrine Immunology and Behavior and Associate Chief, Clinical Neuroendocrinology Branch, National Institute of Mental Health/NIH provided helpful comments and insight on an earlier draft.

References

Ahmed S, Penhale WJ, and Talal N (1985) Sex hormones, immune responses, and autoimmune diseases. *American Journal of Pathology* **121**: 531–51.

Carloni AS (1981) Sex disparities in the distribution of food within rural households. *Food and Nutrition* **7**:3–12.

Chen LC, Huq E, and D'Souza S (1981) Sex bias in the family allocation of food and health care in rural Bangladesh. *Population and Development Review* **7**:55–71.

Fortney JA, Gadalla S, Saleh S, Susanti I, Potts M, and Rogers SM (1987) Causes of death to women of reproductive age in two developing countries. *Population Research and Policy Review* **6**:137–48.

Gittelsohn J (1991) Opening the box: intrahousehold food allocation in rural Nepal. *Social Science and Medicine* **33**:1141–54.

Goodman AH, Lallo J, Armelagos GJ, and Rose JC (1984) Health changes at Dickson Mounds, Illinois (AD 950–1300). In MN Cohen and GJ Armelagos (eds.), *Paleopathology at the Origins of Agriculture*. Orlando: Academic Press, Inc., pp. 271–305.

Grossman C J (1985) Interactions between the gonadal steroids and the immune system. *Science* **227**:257–61.

Guidry DJ (1971) Actinomycosis. In RD Baker (ed.), *Human Infection with Fungi, Actinomycetes and Algae*. New York: Springer-Verlag, pp. 1019–58.

Hall WH and Khan MY (1989) Brucellosis. In PD Hoeprich and MC Jordan (eds.), *Infectious Diseases*, 4th edn. Philadelphia: JB Lippincott Company, pp. 1281–8.

Hoeprich PD (1989) Nonsyphilitic treponematoses. In PD Hoeprich and MC Jordan (eds.), *Infectious Diseases*, 4th edn. Philadelphia: JB Lippincott Company, pp. 1021–34.

Hudson EH (1946) *Treponematosis*. New York: Oxford University Press.

Hurtado AM and Hill KR (1990) Seasonality in a foraging society: variation in diet, work effort, fertility, and sexual division of labor among the Hiwi of Venezuela. *Journal of Anthropological Research* **46**:293–346.

International Union Against Tuberculosis (1964) *Some Statistical Data Concerning Tuberculosis in Europe and North America*. Limoges, France: Imprimerie Bontemps.

Meyers WM (1992) Leprosy. *New Developments in Dermatopathology* **10**:73–96.

Ortner DJ and Putschar WGJ (1981) *Identification of Pathological Conditions in Human Skeletal Remains*. Washington: Smithsonian Institution Press.

Pennington R and Harpending H (1993) *The Structure of an African Pastoralist Community*. Oxford: Clarendon Press.

Powell ML (1988) *Status and Health in Prehistory*. Washington: Smithsonian Institution Press.

Pusey WA (1933) *The History and Epidemiology of Syphilis*. Springfield, Illinois: Charles C. Thomas.

Resnick D and Niwayama G (1988) Osteomyelitis, septic arthritis, and soft tissue infection: the organisms. In D Resnick and G Niwayama (eds.), *Diagnosis of Bone and Joint Disorders*, 2nd edn. Philadelphia: WB Saunders Company, pp. 2524–618.

Roitt IM (1994) *Essential Immunology*. Oxford: Blackwell Scientific Publications.

Rose JC and Hartnady P (1991) Interpretation of infectious skeletal lesions from a historic Afro-American cemetery. In DJ Ortner and AC Aufderheide (eds.),

Human Paleopathology. Washington: Smithsonian Institution Press, pp. 119–27.

Rosenberg EM (1980) Demographic effects of sex-differential nutrition. In NW Jerome, RF Kandel, and GH Pelts (eds.), *Nutritional Anthropology.* New York: Redgrave Publishing Company, pp. 181–203.

Royston E and Lopez AD (1987) On the assessment of maternal mortality. *World Health Statistics Quarterly* **40**:214–24.

Steinbock RT (1976) *Paleopathological Diagnosis and Interpretation.* Springfield, Illinois: Charles C Thomas.

Talal N (1992) Sex hormones and immunity. *In* IM Roitt (ed.), *Encyclopedia of Immunology.* London: Academic Press, pp. 1368–71.

Toure IM (1985) Endemic treponematoses in Togo and other West African states. *Reviews of Infectious Diseases* **7**:S242-S244.

Trueta J (1959) The three types of acute hematogenous osteomyelitis, a clinical and vascular study. *Journal of Bone and Joint Surgery* **41B**:671–80.

Utz JP (1989) Blastomycosis. In PD Hoeprich and MC Jordan (eds.), *Infectious Diseases*, 4th edn. Philadelphia: J.B. Lippincott Company, pp. 510–16.

Weigle KA, Santrich C, Martinez F, Balderrama L, and Saravia NG (1993) Epidemiology of cutaneous leishmaniasis in Colombia: environmental and behavioral risk factors for infection, clinical manifestations, and pathogenicity. *The Journal of Infectious Diseases* **168**:709–14.

Wood JW, Milner GR, Harpending HC, and Weiss KM (1992) The osteological paradox. *Current Anthropology* **33**:343–70.

World Health Organization (1987) *Preventing the Tragedy of Maternal Deaths.* Report on the International Safe Motherhood Conference, Nairobi, Kenya.

World Health Organization Chronicle (1986) Maternal mortality: helping women off the road to death. **40**(5):175–83.

7

Infectious disease, sex, and gender: the complexity of it all

CHARLOTTE A. ROBERTS, MARY E. LEWIS and PHILIP BOOCOCK

The study of sex-related health differences is an important area of research. However, it is also one that is complicated by many factors. Issues of sex and gender, for instance, are important considerations in the assessment of health. The categories of 'male' and 'female' can be viewed as opposing ends of a continuum with considerable overlapping in the middle (Oakley 1985). In addition, the differences between biological sex and gender (the socially determined personal psychological characteristics associated with being male and female), must be recognized (Garrett 1987). Some cultures do not make clear distinctions between categories of 'male' and 'female'. For example, Hollimon (1992) notes the problem of identifying transvestite males (berdaches) in her population, and is unable to consider health differences between the three sex groups. It is likely that this also occurred in the past. This has considerable implications when considering 'male' and 'female' roles, lifestyles, and the development of disease in a society. It cannot be assumed that males and females in the past acted within gender roles similar to those of today.

Another complicating factor is that 'diseases cannot be explained as purely "things in themselves," because they must be analyzed and understood within a human context – that is, in relation to ecology and culture' (Brown *et al.* 1996:183). Culture is instrumental in changing behavior and can lead people to beneficially change their environment in order to enhance health. However, culture can also be maladaptive, directly and indirectly contributing to health problems (Brown *et al.* 1996:184).

From a biocultural perspective, the function and value of males and females in society also affects health. For example, many females in India are discriminated against, which 'starts right from her birth and continues to her last breath' (Bhasin *et al.* 1994:76). For many, this discrimination will affect their predisposition to disease. Researchers must be aware of the many factors that may influence sex differences in disease occurrence. Leprosy, for example, is described as affecting males more frequently than females (Krishnan and Gokam 1992; Sehgal and Chaudhry 1993). There are a number of factors, however, that need to be considered. In India, the detection of female cases of leprosy may be difficult, as women are less likely to come forward for diagnosis because of family commitments, lack of personal resources to attend clinics, and reluctance to expose their body for diagnosis (Butalia 1992). Leatherman (Chapter 8) also points out, in his research on populations in modern Peru and Mexico, the critical need to assess biocultural factors when examining sex differences in health.

Infectious diseases have played a role in human health for millennia. For anthropologists interested in the history and patterns of infectious disease, there are a number of problems inherent in the analysis of skeletal material that can make interpretation of data difficult. For instance, several diseases can potentially affect the skeleton in similar ways and some diseases will leave no mark on the bone. Infectious disease may be acute (rapidly resolved or leading to death) or chronic (a recurring condition that may lead to bone change). In spite of the problem of the 'invisible' nature of many of the infectious diseases, either because they do not manifest themselves in the skeleton or because they kill the person before bone change occurs (Wood *et al.* 1992), the presence of lesions associated with infectious diseases provides useful data on how people adapted to the environments in which they lived.

It is clear that the relationship between the presence of a lesion and a causal agent is not simple. Explanations for the patterns of infectious disease observed in the sexes throughout time can be complicated by the fact that males and females can be predisposed to the same or different factors during their lifetime. In addition, genetic predisposition can contribute to variations in susceptibility to disease by sex. Stini (1985:209) and Ortner (Chapter 6) assert that the female immune response is more effective on average than in males, and that their prognosis for recovery is more favorable. Stinson (1985:141) suggests that males are less buffered against the effects of the environment than females. For example, males have a higher mortality rate in most countries in the first few weeks and months of life. Incidences of respiratory infection are consistently higher in males around the world during the first year of life and over 40 years of age. Reichs (1986) also notes that most of the diseases she examined are more commonly seen in males.

Obviously, determining the variables responsible for health in archaeological populations is not easy. Existing work focusing on health and sex differences is often limited to brief skeletal reports. A few specific studies on sex differences and the occurrence of disease have been published (see Larsen 1986; Hodges 1987; Grauer 1991; Cohen and Bennett 1993; and Grauer and

Roberts 1996). Some of the authors warn that sample size and composition are crucial limitations to the types of questions that can be asked of the data. Lack of associated cultural data can also be limiting. One of the major problems confronting researchers is that the male:female ratio of the sample may not be representative of the original population. Only if one assumes that it is, can sex differences in disease frequency within the sample be treated as a reflection of the total population. However, this is usually hard to determine (Waldron 1994). The fact is, most biological anthropologists working on past populations have to cope with fragmented and incomplete data. Careful consideration of the biases and problems inherent to the study of human skeletal material is thus critical if researchers hope to understand patterns of health and disease in past populations (Wood *et al.* 1992).

The aim of this chapter is to explore the presence of respiratory infectious disease in Medieval populations from England and to determine if sex differences exist. Three questions are addressed:

(1) What role did the environment (urban versus rural) in Medieval England play in the frequency of upper (maxillary sinusitis) and lower (pulmonary) respiratory tract infections?
(2) Were males and females equally susceptible to these infections?
(3) What factors known to cause these conditions may have been operating in these environments?

Infection of the maxillary sinuses

The maxillary sinuses are the largest of the paranasal sinuses. They are lined with a mucous membrane composed of tiny hairs (cilia) that trap pathogens and are instrumental to our body's defense. A blocked ostium (or drainage hole), a change in temperature, or the presence of irritants, may lead to death of the cilia. This renders the sinus dysfunctional and results in infection (Rice 1993). Sinusitis can be acute or chronic, the latter manifesting itself as a recurring and persistent problem. Whether the condition becomes chronic depends on the host's resistance, the virulence of bacilli, and the number of infective organisms. Although maxillary sinusitis has a multifactorial etiology, environmental pollution, overcrowding, poor ventilation, tobacco smoke (Institute for Environmental Health 1994), dental disease (Lundberg 1980), and allergies (Slavin 1982) in the late twentieth century are commonly associated with sinusitis. In addition, other upper respiratory tract infections, and specific infections such as leprosy (Barton 1979) and tuberculosis, may also predispose individuals to the condition. In skeletal populations, chronic infection can be recognized as pitting

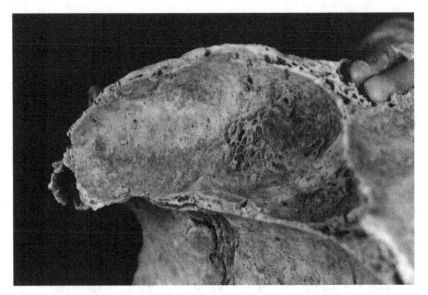

Fig. 7.1 Bone change of sinusitis in maxillary sinus.

and/or new bone formation within the sinus cavity (Boocock *et al.* 1995; Tovi *et al.* 1992) (Figure 7.1).

Infections of the lungs and ribs

Each lung is surrounded by the pleura, a closed sac of serous membrane, with two layers. The pleural cavity contains a thin layer of pleural fluid to prevent friction during breathing. The outer layer of the pleura attaches directly to the internal surface of the ribs (Aiello and Dean 1990). Ribs are instrumental in the act of respiration. Many of the disorders of the respiratory system produce similar signs and symptoms, such as shortness of breath, chest pain, cough, and the production of sputum (Bloom 1975). The main diseases of the upper respiratory tract are the common cold, tonsillitis, tracheitis, laryngitis, sinusitis, and hay fever, while bronchitis, pneumonia, pleurisy, asthma, tuberculosis, cancer, bronchiectasis, and lung abscess, commonly affect the lower respiratory tract. It has been suggested that pulmonary infection can be recognized in skeletal populations by inflammatory lesions of the ribs. These lesions take the form of pitting and/or new bone formation on the internal surfaces of the ribs (Figure 7.2). Roberts *et al.* (1994) and Eyler *et al.* (1996) suggest that tuberculosis is the most likely cause for these lesions, although other pulmonary infections such as pneumonia may be instrumental.

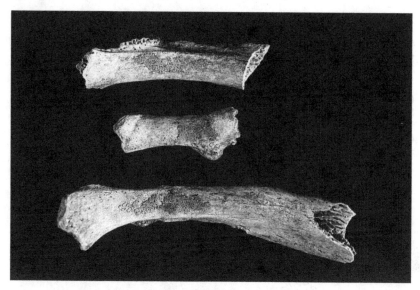

Fig. 7.2 Bone formation on internal surface of rib.

Prevalence rates for maxillary sinusitis and rib lesions are scarce in the published archaeological skeletal record (see Wells 1977; Coenen *et al.* 1995; and Lew and Sirianni 1997, for reports on the presence of sinusitis, and Molto 1990; Sledzik and Bellantoni 1994; and Kelley *et al.* 1994 for reports of rib lesions). Clinical reports on the frequency of these lesions are even more rare (see Eyler *et al.* 1996 for rib lesions). It is suggested that in clinical settings, the bone changes are so subtle in the sinuses and on the ribs that they are not visible on X-ray.

In this study, since all the parts of the respiratory system are anatomically linked and simultaneous occurrence of pulmonary infection and maxillary sinusitis may be expected, the ribs and maxillary sinuses were surveyed. Both conditions appear to be associated more often with urban living and its associated factors (Lewis *et al.* 1995; Chundun 1991).

Materials and methods

Four Medieval skeletal populations from England were selected for this study. Two were rural populations and two were urban (Table 7.1 and Figure 7.3). The populations were selected because of their large numbers and their different environments. The Raunds Furnells population is derived from a rural cemetery in Northamptonshire, dated from the eighth to the twelfth centuries.

Table 7.1. *Sites studied in England*

Site	Century (AD)	Type	Area	Number of individuals
Raunds Furnells	8th–10th	Rural	Midlands	324
St. Helen-on-the-Walls	10th–19th	Urban	North	1042
Wharram Percy	12th–16th	Rural	North	643
Chichester	12th–16th	Urban	South	306

Fig. 7.3 Map of site locations in England.

It includes 324 individuals (Boddington 1996). The urban cemetery of St Helen-on-the-Walls, York, served a poor parish within the city. Its 1042 individuals date from the twelfth to the sixteenth centuries (Dawes and Magilton 1980; Grauer 1991). The Wharram Percy population, with 643 individuals, dates from the tenth to the sixteenth centuries, and was derived from a church cemetery which served several villages in North Yorkshire (Beresford and Hurst 1990). Finally, the hospital cemetery of St. James and St. Mary Magdalene, Chichester, Sussex, associated with a Medieval hospital, yielded 306 individuals. The hospital was established in the twelfth century and was maintained until the sixteenth century (Magilton and Lee 1989). It was founded initially for leprosy sufferers, although later it became a general hospital, providing charity for the sick and poor.

Sex determination of the skeletal material was carried out using methods described in Krogman and İşcan (1986), Bass (1987), and Buikstra and Ubelaker (1994), and concentrated on pelvic features. Age determination relied on the late stages of epiphyseal closure (Bass 1987), dental attrition (Brothwell 1981), sternal rib end degeneration (Loth and İşcan 1989), auricular surface change (Lovejoy *et al.* 1985), and pubic symphyseal degeneration (Brooks and Suchey 1990). Only adult skeletons were included in this study, since one of our aims was to assess differences in males and females for the pathological conditions outlined, and sex estimation of juveniles remains problematic (Saunders 1992).

All adult individuals, attributed a sex, who had at least one maxillary sinus were examined for evidence of maxillary sinusitis (see Boocock *et al.* 1995 for a detailed description of methods and type of pathological changes). New bone formation and/or inflammatory pitting of the ribs (Roberts *et al.* 1994) were recorded only for Raunds Furnells and Chichester because of time constraints. The correlation between dental disease and maxillary sinusitis was also examined; evidence of caries, abscess or periodontal disease associated with sinusitis in the form of an oro-antral fistula was recorded for all skeletons with the maxilla preserved. The differences in frequency rates for sinusitis, rib lesions (where available) and dentally induced sinusitis between the cemeteries, were tested using chi-square analyses. Differences between males and females at each site, and between males for all sites and females for all sites, were also tested using chi-square analyses.

Table 7.2. *Mortality patterns by site, sex, and age at death*

	Raunds Furnells			St. Helen-on-the-Walls			Wharram Percy			Chichester		
	N	n	(%)	N	n	(%)	N	n	(%)	N	n	(%)
Adult males												
17–25	22	16	(11)	36	14	(6)	36	21	(8)	34	23	(17)
26–44	48	35	(23)	53	16	(7)	46	27	(10)	83	29	(22)
45+	24	17	(11)	55	19	(8)	87	49	(18)	11	31	(23)
Adult females												
17–24	29	20	(13)	54	15	(6)	22	14	(5)	15	12	(9)
25–44	25	12	(8)	73	29	(12)	38	27	(10)	50	17	(13)
45+	17	9	(6)	42	21	(9)	53	31	(12)	5	18	(14)
Unknown age/sex												
17+	19	8	(5)	427	74	(30)	85	8	(3)	14	3	(2)
Non-adults												
6–16	140	32	(21)	302	57	(23)	276	91	(34)	94	0	(0)
Total												
	324	149	(46)	1042	245	(24)	643	268	(42)	306	133	(43)

N: Total number of individuals in each sample; *n:* Number of individuals with at least one maxillary sinus preserved; *%:* Percentage of total number of individuals with at least one maxillary sinus preserved.

Results

Removing individuals of unknown age and sex from the sample greatly reduced the numbers available for study. Similarly, an even smaller proportion of the total number of individuals in each population had one maxillary sinus available for examination (Table 7.2). At Raunds Furnells, 149 (46%) individuals out of 324 had one maxillary sinus available for examination. At St. Helen-on-the-Walls, only 245 (24%) individuals were available out of 1042. Wharram Percy yielded 268 (42%) individuals out of 643, and Chichester yielded 133 (43%) out of 306 individuals available for examination. Mortality patterns are also presented in Table 7.2. At all sites, except St Helen-on-the-Walls, more males are present in the populations than females. At no site is the difference between the numbers of males and females in the skeletal sample statistically significant. This pattern is repeated when the numbers of males and females with one maxillary sinus within each population are examined. While the St. Helen-on-the-Walls population had

Table 7.3. *Age distribution of males and females with maxillary sinusitis*

	Raunds Furnells		St. Helen-on-the-Walls		Wharram Percy		Chichester	
	N	*n (%ª)*	*N*	*n (%ª)*	*N*	*n (%ª)*	*N*	*n (%ª)*
All males	68	36 (53)	49	37 (76)	97	43 (44)	84	46 (55)
All females	41	19 (46)	65	45 (69)	72	43 (60)	47	26 (55)
Totals	109	55 (50)	114	82 (72)	169	86 (51)	131	72 (55)
	n	*%ᵇ*	*n*	*%ᵇ*	*n*	*%ᵇ*	*n*	*%ᵇ*
Males (years)								
17–25	8	15	10	12	3	3	10	14
26–44	18	33	14	17	17	19	14	19
45+	10	18	13	16	23	27	22	31
Females (years)								
17–25	9	16	12	15	7	8	8	11
26–44	4	7	22	27	16	19	6	8
45+	6	11	11	13	20	23	12	17
Total	55	100	82	100	86	100	72	100

N: Number of individuals with one maxillary sinus; *n:* Number of individuals with maxillary sinusitis; *%ª*: Percentage of individuals within each sex with maxillary sinusitis from each site; *%ᵇ*: Percentage of individuals with maxillary sinusitis out of total number of individuals with maxillary sinusitis from each site.

more females (*n*=65) with maxillary sinuses than males (*n*=49), at the other sites, more males had the necessary anatomical feature to be included in this study. Statistical comparisons between the proportion of males to females with one maxillary sinus within a population, indicated that the differences were insignificant. Hence, within each population, the number and proportion of individuals with one maxillary sinus could be treated as a reasonable representation of the entire excavated population.

Of all the sites examined, St Helen-on-the-Walls displayed the highest proportion (72%) of individuals with sinusitis (Table 7.3). Chichester had the next highest proportion (55%), followed by Wharram Percy (51%) and Raunds Furnells (50%). Statistically significant differences in the proportion of individuals displaying maxillary sinusitis were found between St. Helen-on-the-Walls (urban) and Wharram Percy (rural) (χ^2=12.54, *p*=0.001). Statistically significant differences were also found between St. Helen-on-the-Walls and Chichester (χ^2=7.53, *p*=0.001), both urban sites, and St. Helen-on-the-Walls and Raunds Furnells (χ^2=10.84, *p*=0.001).

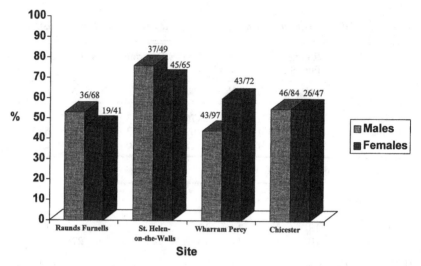

Fig. 7.4 Percentage of individuals (separated according to sex) with maxillary sinusitis out of those individuals with at least one sinus.

When the frequency of maxillary sinusitis was compared between males and females, other patterns emerged (Figure 7.4). At Raunds Furnells, 36 (53%) males and 19 (46%) females displayed the lesions. At St. Helen-on-the-Walls, 37 (76%) males and 45 (69%) females displayed the lesions, while 43 (44%) males and 43 (60%) females at Wharram Percy, and 46 (55%) males and 26 (55%) females at Chichester also suffered from the condition. Only at Wharram Percy were statistically significant differences found (χ^2= 3.92, p=0.05). At Wharram Percy there is also a statistically significant increase in frequency with age in males (χ^2= 11.15, p=0.001). This pattern was not found at any other site.

When exploring dental disease as a possible factor responsible for the development of maxillary sinusitis, several patterns emerge (Table 7.4). At Raunds Furnells, 19 (35%) of 55 individuals with maxillary sinusitis display evidence of the condition being dentally induced. Similarly, 7 (9%) of 82 at St. Helen-on-the-Walls, 23 (27%) of 86 at Wharram Percy and 5 (7%) of 72 at Chichester had evidence of dentally induced sinusitis. It was noticeable that the rural sites of Raunds Furnells and Wharram Percy had more evidence of dentally induced sinusitis than the urban sites. At St. Helen-on-the-Walls, for instance, only 9% of all individuals with maxillary sinusitis had dentally induced lesions. Similarly, at Chichester, only 7% of all cases of maxillary sinusitis indicated that the condition might have been induced dentally. Statistically significant differences were seen in the frequency of dentally induced

Table 7.4. *Age distribution of males and females with dentally induced sinusitis*

	Raunds Furnells			St. Helen-on-the-Walls			Wharram Percy			Chichester		
	N	n	%	N	n	%	N	n	%	N	n	%
Adult males												
17–25	8	4	7	10	1	1	3	1	1	10	0	0
26–44	18	8	15	14	0	0	17	1	1	14	1	1
45+	10	4	7	13	1	1	23	8	9	22	1	1
Adult females												
17–25	9	1	2	12	1	1	7	4	5	8	1	1
26–44	4	1	2	22	3	4	16	9	10	6	1	1
45+	6	1	2	11	1	1	20	0	0	12	1	1
Total	55	19	35	82	7	9	86	23	27	72	5	7

N: Number of individuals with maxillary sinusitis; *n:* Number of individuals with dentally induced sinusitis; *%:* Percentage of individuals within each specific age-at-death category out of all individuals with maxillary sinusitis.

sinusitis between the urban and rural sites. For instance, statistical differences were found between Chichester and Wharram Percy ($\chi^2=10.54$, $p=0.005$), St. Helen-on-the-Walls and Wharram Percy ($\chi^2=9.49$, $p=0.005$), Raunds Furnells and Chichester ($\chi^2=15.5$, $p=0.001$) and Raunds Furnells and St. Helen-on-the-Walls ($\chi^2=11.09$, $p=0.001$).

When examining this evidence by sex and site, a statistically significant higher frequency of dentally induced maxillary sinusitis was only found between males and females at Raunds Furnells ($\chi^2=4.52$, $p=0.05$) (Figure 7.5). When testing for significance of differences of dentally induced sinusitis between the males and the females within and between the sites, there was a statistically significant difference between males at Wharram Percy and St. Helen-on-the-Walls ($\chi^2=4.95$, $p=0.025$), and between males at St. Helen-on-the-Walls and Raunds Furnells ($x^2=14.97$, $p=0.001$)

When the maxillary sinusitis data was linked with the evidence for inflammatory change on the internal rib surfaces, problems developed because the number of individuals with both ribs and sinuses available for examination was small. In addition, from both St. Helen-on-the-Walls and Wharram Percy, data on ribs were unavailable. Only 61 sexed adult individuals at Raunds Furnells, and 28 at Chichester, had both ribs and maxillary sinuses preserved. At Raunds Furnells 5 (4%) of 113 individuals with ribs had lesions, and 2 (3%)

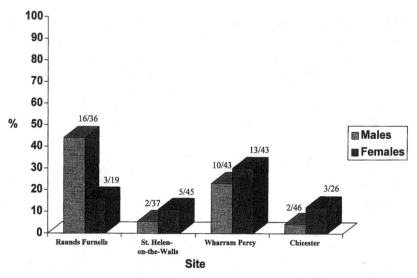

Fig. 7.5 Percentage of individuals (separated according to sex) displaying dentally induced sinusitis out of all those individuals with maxillary sinusitis.

of the 61 sexed adults had both sinuses and ribs involved. At Chichester, 38 (20%) out of 190 individuals with ribs had lesions, and 16 (57%) of 28 sexed adults with ribs and sinuses had both affected; males displayed more rib lesions (23%) than the females (7%). At both sites the differences between males and females were not statistically significant.

Discussion

Problems highlighted in this study included variable skeletal preservation and the omission of a large percentage of each population from the sample because they were non-adults or there was insufficient skeletal data to assign a sex. This left a maximum of 36% (n=114) of the total individuals from St. Helen-on-the-Walls, 66% (n=131) from Chichester, 60% (n=169) from Wharram Percy, and 66% (n=109) from Raunds Furnells available for sex determination. An even smaller percentage of individuals from the sample that was sexed had a maxillary sinus available for examination. In total, only 23% (n=532) of the original 2315 skeletons from the sites could be assigned a sex and examined for the presence of maxillary sinusitis. Samples from the rural sites might be better representations of their entire population, as it is believed that the whole sites were excavated. At the urban sites, only a proportion of the cemeteries were recovered.

In this study, the highest frequencies of maxillary sinusitis were found in the urban sites. No recognizable pattern between males and females in the urban and rural sites was found, although there was a weakly significant difference between the sexes at Wharram Percy (rural). There was a significant difference between the frequency of maxillary sinusitis in males from St. Helen-on-the-Walls (urban) and Wharram Percy. At St. Helen-on-the-Walls and Raunds Furnells (rural), more males were affected. At Wharram Percy, more females were affected, and at Chichester (urban), males and females were equally affected. Significant differences between the urban and rural sites were observed. A weakly significant difference between the frequency of dentally induced sinusitis in males and females was found at Raunds Furnells only.

Many factors could contribute to these results. Some tentative interpretations can be offered, however. The high overall prevalence of maxillary sinusitis in the population studied from York may provide an insight into the urban environment within which they lived. St. Helen-on-the-Walls was one of the poorest parishes in the Medieval city of York. Archaeological evidence indicates that a foundry, tannery, brewery, and apothecary were located very close by in the Bedern parish. The by-products of tanning and founding are irritant particles, which are released into the atmosphere and are known to be associated with lung infections and sinusitis. In addition, the thirteenth-century establishment of a lime kiln to provide materials for the construction of York Minster would have meant burning large amounts of sea coal, which produces smoke and sulphur dioxide, both significant pollutants (Brimblecombe 1976). Coal burning in individual houses also contributed to local and general environmental pollution, predisposing people to respiratory disease. Ramazzini (1633–1714), the father of occupational medicine, recognized the link between coal and respiratory disease (Blanc and Nadel 1994).

The urban site of Chichester, it is assumed, had a similar set of urban pollutants, although prevalence rates for maxillary sinusitis for this site were lower than for St. Helen-on-the-Walls. The nature of this site, however, must be taken into consideration. Founded as a hospital for leprosy sufferers, it is probable that the patients were segregated from the rest of Chichester and perhaps were less exposed to environmental pollutants. Another factor to consider is that leprosy can predispose individuals to maxillary sinusitis (Barton 1979; Hauhnar *et al.* 1992), consequently involving the respiratory system (Kaur *et al.* 1978). In an independent study, however, Boocock *et al.* (1995) found no difference in the occurrence of maxillary sinusitis between leprous and non-leprous skeletons. It is, therefore, assumed that leprosy did not complicate the picture in this case.

The lower frequencies of maxillary sinusitis within the populations from the rural sites suggests that factors other than those occurring in urban environments were operating to induce sinusitis. Certainly, the higher frequency of dentally induced sinusitis suggests that dental disease was a more important etiological factor at the rural sites than at St. Helen-on-the-Walls or at Chichester. While specific dietary information for the populations examined in this study is unavailable, within historic British populations it has been reported that dental disease, particularly caries and antemortem tooth loss, was common in later and post-Medieval groups (Moore and Corbett 1973). Sucrose, a recognized factor in cariogenic tooth decay, was rarely a part of the British diet until after the twelfth century. Even so, at Raunds Furnells, presumably before refined sugar was widely available, the frequency of dentally induced sinusitis is high. Cereal crops, however, are known to contain fermentable carbohydrates, necessary for caries to develop. Hence, it is likely that dental caries would develop in the agricultural populations studied. Agricultural villages, such as Raunds Furnells and Wharram Percy, were also environments containing many pathogenic microorganisms such as fungal spores and pollen, dust from the processing of crops, molds, zoonotic disease, and parasitic infection (Figure 7.6). A recent study by McCurdy *et al.* (1996) suggests that organic and inorganic dust and smoke produced by burning agricultural waste leads to chronic bronchitis and asthma in rice farmers in the late twentieth century. All these factors are relevant to rural-based maxillary sinusitis.

Fig. 7.6 Wharram Percy Medieval Village. Reconstruction of Medieval peasants doing agricultural activities. By Peter Dunn (with permission of English Heritage Photographic Library).

General environmental pollution in the urban groups and dental disease in the rural populations seem to be the most likely causes of the sinusitis prevalence observed. However, other factors such as poor hygiene, inadequate diet and an inferior water supply could have been instrumental in compromising immune systems and making people more susceptible to these conditions. Similarly, indoor pollution from wood burning can predispose people to respiratory disease. Cleary and Blackburn (1968), for example, found a high prevalence of chronic lung disease in New Guinea Highlanders who occupy huts which have a high concentration of smoke from wood fires that were tended for most of the day and night. Master (1974) also found in New Guinea that pulmonary disease was the most important cause of morbidity and mortality, and was correlated with increasing age. Males and females were equally affected. However, studies of other cultures have found males and females to be differentially affected (Rajpandey 1984a, 1984b).

A wide variety of respiratory diseases are associated with elevated particulate pollution in houses using unprocessed biomass fuels such as dung (Albalak 1997). It is likely that this was the case in the past, and that fires in poorly ventilated houses contributed to upper and lower respiratory tract infection in both urban and rural sites. Little is known of the housing at the sites examined in this study, although it is known that *in rural areas, long houses* where both humans and animals co-habited, contained a central hearth (Figure 7.7). It is also believed that housing would have been poorly ventilated (as

Fig. 7.7 Wharram Percy Medieval Village. Reconstruction of interior of 13th century cruck built long-house. By Peter Dunn (with permission of English Heritage Photographic Library).

in many developing countries today), allowing smoke from fires to accumulate quickly and exposing the occupants to harmful particulate matter.

Although the differences in frequency rates of maxillary sinusitis between males and females was found to be statistically insignificant, factors must explain the *actual* differences seen at each site. Ethnographic studies demonstrate complexity and cross-cultural variability in gender arrangements in living societies (Conkey and Spector 1984:14), although the androcentrism in ethnography often restricts the use of these data in studies such as this. In addition, growth of a class society in the Medieval Period in England, and political and economic factors may all have had their part to play in the patterning of health in males and females, where timing and exposure to risk for the sexes will have been affected (Reiter 1975; Schell 1997). The World Health Organization certainly recognizes this and describes how it sees disparities in morbidity and mortality between the sexes as critical indicators of equity in society (Smyke 1991; Miles 1991). However, in this study the differences seen in the health indicator observed do not seem to reflect major differences in exposure to risk, but rather general similarities in exposure with minor differences, perhaps reflecting occupation and lifestyle patterns. Interestingly, however, Miles (1991) cites a study by Leviatan and Cohen conducted in 1985, where kibbutz and non-kibbutz population life expectancy was examined and where the intergender difference was less in the former; this was explained by the greater similarity of life experiences and social roles of males and females in a kibbutz compared to a non-kibbutz group. This stresses the need, when comparing sex differences in health, to consider the cultural context from which the sample is derived, and the many factors influencing patterns.

Conclusion

The aims of this study were to consider sex differences in respiratory disease. Three questions were posed: (1) were people living in Medieval urban environments more likely to contract respiratory tract infections than those living in rural conditions; (2) were males more susceptible to respiratory disease than females; and (3) what might be the etiology of the infections?

This study has highlighted a number of factors predisposing males and females to maxillary sinusitis and other related conditions. The differences in urban and rural environments in terms of the level of pollution present (both local and general), climate, standards of hygiene, housing conditions, socio-economic status, population density, access to a balanced diet, and occupations practiced can be significant in the late twentieth century. Males and females may be differentially exposed to these factors and be more or less susceptible to disease processes depending on their individual behavioral pattern. Males and females may also be immunologically different. This study showed that people living in both urban and rural environments were susceptible to developing respiratory disease but that urban groups had higher prevalence rates. Differences between urban and rural rates were highly significant in all cases. Only at Wharram Percy was there a significant difference between male and female prevalence. This suggests that both sexes were exposed to similar predisposing factors for respiratory disease. When considering dental disease as a factor for maxillary sinusitis development, the rural groups had higher prevalence rates of dentally induced sinusitis, and the differences between the urban and rural groups were highly significant. Differences between the sexes within the sites showed a much higher frequency in males at Raunds Furnells only.

Apart from the evidence of dental disease, there was no clear evidence for the specific causes of the sinusitis observed. General environmental pollution at the urban sites probably caused the majority of the sinusitis seen, i.e., there was less frequent occurrence of sinusitis and underlying dental disease, although it is likely that smokey home environments may have contributed considerably to the development of respiratory disease. Any genetic predisposition to developing this non-specific infectious disease would have been overshadowed by other factors.

This study has considered male and female susceptibility to a non-specific infectious disease in urban and rural environments. It is hoped that further work will expand on the particular reasons why males and females develop respiratory tract infections and will try to separate out those factors which may be specific to males and females (e.g., see Ortner, Chapter 6, for discussion on sex differences in immunological reactivity). Using a holistic biocultural approach to study large populations, alongside studies of living traditional populations, may help to clarify this fascinating area of sex differences in health. However, 'the picture will be extremely complex [and] the components of the environment are so numerous and varied that it will be very difficult to identify the cause of chronic and low grade illnesses ... [however] the insights provided by the study of human biology can contribute substantially to our understanding' (Boyden 1972:168). Obviously, the study of sex-related differences in health produces more questions than it answers, although, as Slocum (1975:36) notes, 'the basis of any discipline is not the answers one gets but the questions it asks.'

Acknowledgments

The authors would like to thank the following organizations and people for access to skeletal collections and data: Northamptonshire Archaeology Unit and Faye Powell (Raunds Furnells), Frances Lee, Zoe Chundun and Chichester District Council (Chichester), Peter Addyman, York Archaeological Trust, Elizabeth Hartley and the Yorkshire Museum, and Anne Grauer (St. Helen-on-the-Walls), Simon Mays and English Heritage (Wharram Percy), and the Wellcome Trust, Science and Engineering Research Council, Natural Environmental Research Council and Nuffield Foundation for funding some of this research. Jean Brown, photographic technician of the Department of Archaeological Sciences, reproduced Figures 7.2 and 7.3, and Celia Harding of English Heritage Photograph Library arranged reproduction of Figures 7.6 and 7.7.

References

Aiello L and Dean C (1990) *An Introduction to Human Evolutionary Anatomy*. London: Academic Press.

Albalak R (1997) Indoor air pollution in rural areas of the developing world. *American Journal of Physical Anthropology*, Supplement 24: 64.

Barton RPE (1979) Radiological changes of the paranasal sinuses in leprosy. *Journal of Laryngology and Otology* **93**:597–600.

Bass WM (1987*) Human Osteology. A Field Guide and Manual*. Missouri: Missouri Archaeological Society Special Publication 2.

Beresford M and Hurst J (1990) *The English Heritage Book of Wharram Percy Deserted Medieval Village*. London: English Heritage and Batsford.

Bhasin MK, Walter H, and Danker-Hopfe H (1994) *People of India. An Investigation of Biological Variability in Ecological, Ethno-economic and Linguistic Groups*. Delhi: Kamla-Raj Enterprises.

Blanc PD and Nadel JA (1994): Clearing the air: the links between occupational and environmental air pollution control. *Public Health Review* **22**:251–70.

Bloom A (1975) *Toohey's Medicine for Nurses*, 11th edn. Edinburgh: Churchill Livingstone.

Boddington A (1996) *Raunds Furnells: The Anglo-Saxon Church and Churchyard*. London: English Heritage.

Boocock P, Roberts CA, and Manchester K (1995) Maxillary sinusitis in Medieval Chichester, England. *American Journal of Physical Anthropology* **98**(4):483–95.

Boyden S (1972) Biological views of problems of urban health. *Human Biology in Oceania* **1**(3):159–69.

Brimblecombe P (1976) Attitudes and responses towards air pollution in Medieval England. *Journal of the Air Pollution Control Association* **28**(2):115–18.

Brooks ST and Suchey JM (1990) Skeletal age determination based on the os pubis: a comparison of the Ascadi–Nemeskeri and Suchey–Brooks methods. *Human Evolution* **5**:227–38.

Brothwell D (1981) *Digging Up Bones*. London: British Museum (Natural History).

Brown PJ, Inhorn MC, and Smith DJ (1996) Disease, ecology and human behaviour. In CF Sargent and TM Johnson (eds.), *Medical Anthropology. Contemporary Theory and Method.* Revised Edition. London: Praeger, pp. 183–218.

Buikstra JE and Ubelaker DH (eds.) (1994) *Standards for Data Collection from Human Skeletal Remains.* Proceedings of a Seminar at the Field Museum of Natural History. Fayetteville: Arkansas Archeological Survey Research Series No. 44.

Butalia U (1992) *The Story Within the Story. Women Fight Against Leprosy.* New Delhi: Danlep.

Chundun Z (1991) *The Significance of Rib Lesions in Individuals from a Chichester Medieval Hospital.* MSc thesis, Calvin Wells Laboratory, Department of Archaeological Sciences, University of Bradford.

Cleary GJ and Blackburn CRB (1968) Air pollution in native huts in the highlands of New Guinea. *Archives of Environmental Health* 17:785–94.

Coenen V, Bruintjes TJD, and Panhuysen RGAM (1995) Maxillary sinusitis in Medieval Maastricht, The Netherlands. *Journal of Paleopathology* 7(2):90.

Cohen MN and Bennett S (1993) Skeletal evidence for sex roles and gender hierarchies in prehistory. In BD Miller (ed.), *Sex and Gender Hierarchies.* New York: Cambridge University Press, pp. 273–98.

Conkey MW and Spector J (1984) Archaeology and the study of gender. In M Schiffer (ed.), *Advances in Archaeological Method and Theory*, vol. 7. New York: Academic Press, pp. 1–38.

Dawes J and Magilton J (1980) *The Cemetery of St Helen-on-the-Walls. The Archaeology of York. The Medieval Cemeteries* 12/1. London: Council for British Archaeology for York Archaeological Trust.

Eyler WR, Monsein LH, Beute GH, Tilley B, Schultz LR, and Schmitt WGH (1996) Rib enlargement in patients with chronic pleural disease. *American Journal of Radiology* 167:921–6.

Garrett S (1987) *Gender.* London: Tavistock Publications.

Grauer AL (1991) Life patterns of women from Medieval York. In D Walde and ND Willows (eds.), *Archaeology of Gender.* Calgary: Archaeological Association of the University of Calgary. Proceedings of the 22nd Chacmool Conference, pp. 407–13.

Grauer AL and Roberts CA (1996) Paleoepidemiology, healing and possible treatment of trauma in the Medieval cemetery population of St. Helen-on-the-Walls, York, England. *American Journal of Physical Anthropology* 100:531–44.

Hauhnar CZ, Mann SBS, Sharma VK, Kaur S, Mehta S, and Radotra BD (1992) Maxillary antrum involvement in multibacillary leprosy: a radiologic, sinuscopic and histologic assessment. *International Journal of Leprosy* 60(3):390–5.

Hodges DC (1987) Health and agricultural intensification in the prehistoric valley of Oaxaca, Mexico. *American Journal of Physical Anthropology* 73:323–32.

Hollimon SE (1992) Health consequences of sexual division of labor among native Americans: the Chumash of California and the Arikara of the Northern Plains. In C Claassen (ed.), *Exploring Gender Through Archaeology: Selected Papers from the 1991 Boone Conference.* Monographs in World Archaeology No. 11. Wisconsin: Prehistory Press, pp. 81–88.

Institute for Environmental Health (1994) *Report on Air Pollution and Respiratory Disease: UK Research Priorities.* Medical Research Council Report R2.

Kaur S, Malik SK, Kumar B, Singh MP, and Chakravarty RN (1978) Respiratory system involvement in leprosy. *International Journal of Leprosy* 47(1):18–25.

Kelley MA, Murphy SP, Levesque DR, and Sledzik P (1994) Respiratory disease among protohistoric and early historic Plains Indians. In DW Owsley and RL Jantz (eds.), *Skeletal Biology in the Great Plains. Migration, Warfare, Health and Subsistence*. Washington DC: Smithsonian Institution Press, pp. 123–30.

Krishnan BK and Gokam A (1992) Study of leprosy among slum dwellers in Pune. Part 1: Prevalence. *Indian Journal of Public Health* **36**(3):78–86.

Krogman WM and İşcan MY (1986) *The Human Skeleton in Forensic Medicine*. Illinois: Charles Thomas.

Larsen CS (1986) Health and disease in prehistoric Georgia: the transition to agriculture. In MN Cohen and GJ Armelagos (eds.), *Paleopathology at the Origins of Agriculture*. London: Academic Press, pp. 367–92.

Leviatan U and Cohen J (1985) Gender differences in life expectancy among kibbutz members. *Social Science and Medicine* **21**(5):545–51.

Lew R and Sirianni JE (1997) Incidence of maxillary sinus infection in Highland Park cemetery. *American Journal of Physical Anthropology*, Supplement 24:154.

Lewis ME, Roberts CA, and Manchester K (1995) A comparative study of the prevalence of maxillary sinusitis in Medieval urban and rural populations in Northern England. *American Journal of Physical Anthropology* **98**(4):497– 506.

Loth S and İşcan MY (1989) Morphological assessment of age in the adult: the thoracic region. In MY Iscan (ed.), *Age Markers in the Human Skeleton*. Springfield: Charles Thomas, pp. 105–35.

Lovejoy CO, Meindl RS, Pryzbeck TR, and Mensforth RP (1985) Chronological metamorphosis of the auricular surface of the ilium: a new method for the determination of adult skeletal age. *American Journal of Physical Anthropology* **68**:15–28.

Lundberg C (1980) Dental sinusitis. *Swedish Dental Journal* 4:63–7.

Magilton JR and Lee F (1989) The leper hospital of St James and St Mary Magdalene, Chichester. In CA Roberts, F Lee, and J Bintliff (eds.), *Burial Archaeology: Current Research, Methods and Developments*. British Archaeological Reports British Series 211, Oxford, pp. 249–65.

Master KM (1974) Air pollution in New Guinea. Cause of chronic pulmonary disease among Stone-Age natives in the Highlands. *Journal of the American Medical Association* **228**(13):1653–5.

McCurdy SA, Ferguson TJ, Goldsmith DF, Parker JE, and Schenker MB (1996) Respiratory health of California rice farmers. *American Journal of Respiratory Critical Care Medicine* **153**:1553–9.

Miles A (1991) *Women, Health and Medicine*. London: Open University Press.

Molto JE (1990) Differential diagnosis of rib lesions: a case study from Middle Woodland Southern Ontario, circa 230AD. *American Journal of Physical Anthropology* **83**:439–47.

Moore WJ and Corbett E (1973) Distribution of dental caries in ancient British populations. *Caries Research* **7**:139–53.

Oakley A (1985) *Sex, Gender and Society*. Aldershot: Gower Publishing Company Limited.

Rajpandey M (1984a) Prevalence of chronic bronchitis in a rural community of the Hill Region of Nepal. *Thorax* **39**:331–6.

Rajpandey M (1984b) Domestic smoke pollution and chronic bronchitis in a rural community of the Hill Region of Nepal. *Thorax* **39**:337–9.

Reichs K (1986) Forensic implications of skeletal pathology: sex. In K Reichs (ed.), *Forensic Osteology. Advances in the Identification of Human Remains.* Springfield: Charles Thomas, pp.112–42.

Reiter RR (Ed.) (1975) *Toward an Anthropology of Women.* London: Monthly Review Press.

Rice DH (1993) Inflammatory diseases of the sinuses. *Otolaryngologic Clinics of North America* **26**(4):619–22.

Roberts CA, Lucy D, and Manchester K (1994) Inflammatory lesions of ribs: analysis of the Terry Collection. *American Journal of Physical Anthropology* **94**(2):169–82.

Saunders S (1992) Subadult skeletons and growth related studies. In SR Saunders and MA Katzenberg (eds.), *Skeletal Biology of Past Peoples: Research Methods.* New York: Wiley-Liss, pp. 1–20.

Schell LM (1997) Culture as a stressor: a revised model of biocultural interaction. *American Journal of Physical Anthropology* **102**:62–77.

Sehgal VN and Chaudhry AK (1993) Leprosy in children. A prospective study. *International Journal of Dermatology* **32**(3):194–7.

Slavin RG (1982) Relationship of nasal disease and sinusitis to bronchial asthma. *Annals of Allergy* **49**:76–80.

Sledzik P and Bellantoni N (1994) Brief communication: bioarcheological and biocultural evidence for the New England vampire folk belief. *American Journal of Physical Anthropology* **94**:269–74.

Slocum S (1975) Woman the gatherer: male bias in anthropology. In RR Reiter (ed.), *Toward an Anthropology of Women.* London: Monthly Review Press, pp. 36–50.

Smyke P (1991) *Women and Health.* London: Zed Books Ltd.

Stini W (1985) Growth rates and sexual dimorphism. In RI Gilbert and JH Mielke (eds.), *Analysis of Prehistoric Diets.* New York: Academic Press, pp. 191–226.

Stinson S (1985) Sex differences in environmental sensitivity during growth and development. *Yearbook of Physical Anthropology* **28**:123–47.

Tovi F, Benharroch D, Garot A, and Hertzanu Y (1992) Osteoblastic osteitis of the maxillary sinus. *Laryngoscope* **102**:427–30.

Waldron T (1994) *Counting the Dead. The Epidemiology of Skeletal Populations.* Chichester: John Wiley and Sons.

Wells C (1977) Disease of the maxillary sinus in antiquity. *Medical and Biological Illustration* **27**:173–8.

Wood JW, Milner GR, Harpending HC, and Weiss KM (1992) The osteological paradox. Problems of inferring health from skeletal samples. *Current Anthropology* **33**(4):343–70.

8

Gender differences in health and illness among rural populations in Latin America

THOMAS L. LEATHERMAN

Within the local and regional contexts of global economic transformations, women in Latin America and throughout the developing world face inequalities in status, wealth, and power. Development efforts aimed at capitalizing rural economies have tended to neglect women's critical role in household economies, and to devalue their efforts in monetized economies. Women are increasingly operating in spheres of production, household and family maintenance, and biological reproduction (Beneria 1979). These multiple roles are recognized in the 'double day' of work that many women must perform (i.e., a full-days work outside the home plus a full-day completing domestic chores). A demographic transition toward fewer births, often expected to follow 'modernization', has not universally occurred, and levels of fertility and maternal and infant mortality remain alarmingly high throughout much of the developing world. Together, these changing realities have heightened the vulnerability of women to the biological costs of inequality (e.g., see Harrington 1983; Browner 1989; Leslie and Paolisso 1989).

Both feminist and development literatures have documented how women's roles in social and biological reproduction can structure their productive activities, and how production activities might influence reproductive behavior (Beneria and Roldan 1987). Yet, researchers have paid less attention to the ways in which these interactions may affect women's health (Browner 1989:461). Also, while health research has focused on the myriad of health problems women face, particularly in contexts of rural poverty, relatively few studies have examined women's health in the household context (Browner 1989). Yet, the household is a crucial site of processes linking production to social and generational (biological) reproduction (Deere 1990). In order to examine whether women are disproportionately ill, and the degree to which their structural position in global, regional, and local economies may affect inequalities in health, it is necessary to examine gender differences in health in the household context.

In this chapter, I examine gender differences in health in two regions of Latin America the Andes of southern Peru and the Mayan lowlands of the Yucatan Peninsula. The Andean data are based on field work carried out in 1983–84 to assess the consequences and responses to illness among small-scale farmers and herders in the District of Nunoa in the southern Peruvian Andes (Thomas *et al.* 1988). The Mayan case comes from more recent research in 1989 and 1991 on household health and economy in the community of Yalcoba, in the Yucatan of Mexico, but with a more explicit focus on women's health in the context of male out-migration to work in the tourist economy. In both cases, I place the analysis of gender differences in illness within the household division of labor, and in the macro-economic processes that have altered local economic and social relations of production and repro-duction. I suggest that where the household division of labor is most distinct, and men and women's economic roles and power are most disparate, gender differences in illness are greater. Also, where poor economic conditions dictate marginal diets and heavy work loads, high levels of fertility might importantly contribute to high levels of chronic illness in adult women.

The case studies and discussion presented in this chapter provide an example of contemporaneous biocultural studies of health that have implica-tions for gendered health in history and prehistory. Gender-based differences in a division of labor, access to resources, and political, social, and economic power are important to our understanding of both past and present societies. The health status of populations and hierarchies of health within populations are important reflections of living conditions, social relations, and everyday realities of human societies regardless of the era. Indeed, the reflection of social status differences in health profiles based on pathological indicators of stress and disease is a well established research agenda in skeletal biology (Goodman 1998; Martin 1998; see also Grauer *et al.*, Chapter 10 for a dis-cussion of the effects of engendered social roles on disease patterns in a historic population; Storey, Chapter 9 for discussions of women's health in prehistory). Status–health relationships must be approached critically (see Saitta 1998) in order to yield important and useful information for improved understandings of both historic and prehistoric health *and* social relations. Thus a central objective of the contemporaneous cases presented in this chapter is to provide data and perspectives which might help open new avenues of inquiry into gendered social relations and health in the past and in the present.

Contexts of women's work and health

Throughout rural Latin America, agrarian transformations have occurred in response to a host of factors including land reform and capital penetration into rural sectors. These transformations have important implications for health (see Leatherman and Gordon 1994). The picture that emerges is one that has been played out in many contexts in the developing world. Capitalist penetration into rural economies has stimulated the commoditization of goods and labor and has affected a change in household production strategies toward a mix of home food production and wage labor (i.e., semi-proletarianization). Neither strategy alone is sufficient to meet basic household needs (de Janvry 1981). Women increasingly enter the work force, though often are limited to participating in the informal economy of piecework and petty commerce – weaving others' hammocks, spinning others' wool, and marketing others' goods. In rural Latin America few wage-earning opportunities are available to women, they regularly receive less pay than men for the same work, and in some cases are even prevented from owning land. Men work at wage jobs locally, and/or migrate to urban centers in search of work, leaving women to maintain family and home until they return, and in some cases to carry out all agricultural tasks as well as informal cash-earning activities. As men stay away from home for longer periods, fail to provide cash for household and family needs, or never return, an increasing number of rural households are headed by single females on a permanent or part-time basis. In this case, women are the primary producers as well as being responsible for the daily maintenance of family and household needs. Indeed, time allocation studies in rural communities throughout the developing world show that women work long days, often longer than men, and increasingly they work more 'double days' as they intensify their inputs into market work as well as home food production (Browner and Leslie 1995). Many women spend anywhere from 10 to 16 hours a day in housework, food preparation, child care, and nursing sick relatives, as well as in generating products and income for meeting household basic needs (Browner and Leslie 1995). In addition, mean completed fertility rates are often high, and women spend much of their early adult life pregnant, lactating, and caring for small children. These reproduction processes can be physically stressful, especially in the context of poor nutrition.

These patterns have led to a concern over women's health in contexts of high fertility and marginal living conditions (e.g., Harrington 1983; Daltabuit 1988; Browner and Leslie 1995). As the effects of high fertility, poor nutrition and heavy work loads are heightened, one can expect women's health to suffer. Moreover, the poor health and nutritional status of women (and men) affect their ability to effectively carry out their multiple roles in household production and reproduction, and thus can serve to perpetuate poverty and poor health (cf. Leatherman 1996). As summarized by Browner and Leslie (1995:262):

Women's health and nutritional status influences, and is in turn influenced by, these multiple roles in a number of important ways. With regards to household production, a woman's health and nutritional status directly influence her ability to conceive, give birth and breast-feed, as well as her infant's health at birth and nutritional status. With regard to market production, a woman's physical capacity to produce food and/or generate income is directly affected by her own health and nutritional status. Competing demands on a woman's time on household versus market production may constrain her ability to protect and promote her own and her family's health. For some poor women, the demands of their multiple roles require more energy than their food intake provides, leading to further deteriorating health and malnutrition.

These relationships and concerns noted by Harrington (1983), Daltabuit (1988), and Browner and Leslie (1995) on women's health are evident in the regions of southern Peru and eastern Mexico from which both case studies are drawn. Both are sites of agrarian economies undergoing fairly rapid transformations, due in large part, to capital penetration and the commoditization of goods and labor. In Peru, change is linked to an agrarian reform, market expansion, and monetization of the rural economy as part of an attempt to promote urban industrial development (Painter 1984; Leatherman 1996). In the Yucatan, change has occurred rapidly due to tourism-led economic development (Daltabuit and Pi-Sunyer 1990). Both geographical areas are composed of poor rural producers with limited availability of land, poorly developed infrastructure, and limited permanent wage work opportunities. Both areas have undergone capitalist development which has drawn rural farming households into the market economy primarily as wage workers. While women work in the market economy, most wage work is performed by men. Women are increasingly left to perform the tasks of daily household production, in addition to work in agro-pastoral production and piecework in the informal economy. Time allocation studies in both locales showed women to work longer hours than men (Leatherman 1987; Daltabuit 1988). Reproduction surveys found that women in both locales had high levels of fertility – between seven and eight births. Dietary and anthropometric surveys found diet and nutritional status to be marginal in each community studied (Daltabuit 1988; Leonard and Thomas 1988). Finally, both locations have limited infrastructures of environmental quality/hygiene; some running water but no sewage disposal facilities. State run health clinics provided limited biomedical services in both locales, and local health systems reflected a syncretism of indigenous and biomedical perspectives on the etiology and treatment of illness.

The Andean context of household relations

Household economies in the southern Andes have been historically based on agro-pastoral production. A relatively high degree of overlap in production tasks are reported for men and women. Women and children, however, perform most herding tasks. Women also perform most household work (food preparation, child care, collecting water, etc.). Men do the majority of the heavier work (e.g., plowing) in agriculture. Men in rural farming–herding communities also spin wool and weave cloth, and carry out some of the basic household maintenance tasks.

Over the past two decades the combined effects of agrarian reform policies, the growth of markets, and the commoditization of goods and labor, have led to transformations in the economy and social relations of Andean households. An agrarian reform was instituted in the early 1970s in an effort to redress inequities in the land base among the rural peasantry. In effect, reform initiatives replaced haciendas with even larger cooperatives that employed fewer people. As a result, the urban town in the district almost doubled in size due, in part, to an increased influx of landless households from the surrounding countryside. Also, the need for cash increased to pay helpers during planting and harvest, and for the purchase of basics like food and clothing. But the irregular availability of wage work and extremely low wages (about US $1 for a day's work) meant that most wage incomes were insufficient to meet these basic needs; this underscored the importance of farming (Leatherman 1996). Thus, like elsewhere in Latin America (de Janvry 1981), a large semi-proletarian class emerged, which was unable to fully meet its basic needs through farming and herding, or through wage work.

As Nunoan households became more involved in the market economy, the division of labor in the household became more stratified. In general, the more the household was involved in wage work, the greater the gender based division of labor. This was most apparent in the one semi-urban town in which we worked, where 20% of the households were landless and those with land regularly supplemented farming production with up to five different sorts of cash earning activities. It was common for men to be engaged in wage work of a non-agricultural nature and out of the home, and for women to do more piecework (e.g., spinning and weaving) and small scale mercantile activities (e.g., selling others' goods in the local market) as well as the majority of household work. Time allocation studies from the town showed that women did less farming work but spent almost as much time as men in market activities and twice the time in household and family maintenance tasks such as hauling water and wood, food preparation, child care, and washing clothes (Leatherman 1987, 1992). Also, while adolescent and adult males in both rural and semi-urban communities migrated on a regular or part-time basis to work in wage labor, men in the semi-urban town migrated more frequently and for longer periods, in part because of less access to land for farming. As a result, over 20% of the town households were effectively single woman-headed households, where women's work loads were obviously higher and household incomes were usually lower.

The Mayan context of household relations

A generation ago, the southeastern quadrant of the Yucatan peninsula remained one of the most inaccessible locations in Mexico. Today, the Maya of the eastern Yucatan find themselves part of a rapidly developing tourist economy. In the early 1970s, Cancun was an isolated fishing village with 426 inhabitants, but in two decades it had become the most important city in the region with a population of approximately 400,000 people and an annual growth rate of 20%.

The highly capital intensive development of Cancun and the coast created a labor market based largely in construction and in service industries to tourists (Daltabuit and Pi-Sunyer 1990). Even in communities from the interior, ones that tourists rarely see, large numbers of men migrate for work to Cancun and to the coast on a weekly or more permanent basis. For established families, this places an added burden on women who must maintain home and family in their husbands' absence.

Yalcoba is a Mayan village of about 1450 inhabitants located in a Maize Production Zone in the interior of the Yucatan Peninsula approximately 100 miles from Cancun. It is a community of contrasts. A typical house of *bajareque* construction (pole and thatch) has a television in a prominent place. There is no good source of clean water, waste disposal, or sanitation in the village, but there is cable TV. The majority of the inhabitants ($\pm75\%$) practice slash and burn agriculture, but wage work is now the major source of household income.

The transformation of the subsistence economy into a cash economy has been reinforced by limited access to *ejido* lands (i.e., communal agricultural land passed down though families, but until 1992 these lands could not be sold), making traditional subsistence strategies less viable. Also, increased consumption norms and the commoditization of basic needs, including food, make a steady source of cash income essential to daily household production. In 1991, between 60 and 75% of the households in Yalcoba reported wage work in or near Cancun as a major source of household income.

As men migrate into Cancun, women are left in the village to attend to local production and household reproduction activities. Men that work in Cancun return to Yalcoba on the weekends, and some return only every two weeks. The situation is often exacerbated by the fact that men may not find work for one or more weeks, or may spend the majority of their earnings before returning home for the weekend.

Time allocation data collected in 1989 illustrated that women worked longer hours and with fewer rests than men, confirming an earlier and more extensive survey by Daltabuit in 1986. Women are in charge of virtually all domestic work and increasingly spend their time in the relatively few available income generating activities – none of which provides a sizeable or steady income. Thus, in the context of tourism-led economic development, these women are increasingly operating in spheres of production and household production, and with relatively little control over household budgets (Daltabuit 1988).

Methodology

Both of the case-studies described below are based on research designs aimed at elucidating relationships between health and household economies. Data on household health were collected with structured and semi-structured illness recall interviews. For the most part, female heads of household (and adult males who were present) were interviewed about the health of household members. One set of questions asked about illness episodes among all house-hold members over two-week and one-month periods, and about occurrences of more serious illness capable of disrupting patterns of work (measured as days of work lost over the same recall periods). Structured illness symptoma-tologies were collected for all household members in Nunoa and for adult women in Yalcoba (see Carey 1990 and Leatherman 1987, 1996 for a more detailed discussion of methodologies for assessing health status). Anthropo-metric surveys were carried out among school-aged children to provide a general assessment of community nutrition and health (Daltabuit 1988; Leatherman *et al.* 1995).

Information on household farming and wage production activities, and on access to land and other production inputs, was collected along with a material resource inventory to assess economic status and production strategies. Time allocation surveys including 'spot check', recall, and continuous observations provided information on the time spent in different activities by household members, and is used here along with the economic surveys to assess division of labor and work loads in the households (see Daltabuit 1988 and Leatherman 1987, 1992 for more detailed methodologies).

In addition, assessments of health, work, household economy, and reproduction surveys of adult women were done in both locales. The high levels of fertility indicated from these surveys provide a measure of reproductive load for mothers in both communities. For women in Yalcoba, data from these surveys also were used to construct a reproductive stress index modified from similar indices calculated by Daltabuit (1988) following the equation of Harrington (1983:135). Harrington calculated an index of physical and nutritional stress defined as the 'proportion of a woman's reproductive life to date spent either pregnant or breast-feeding' – reproductive life was defined as time between the first pregnancy and the present. The reproductive stress index used here defined reproductive life as extending from age 15 to the present, but not exceeding a maximum period of 30 years (between the ages of 15 and 45). Young women in Yalcoba marry and occasionally give birth by age 15, and no women in the sample had given birth past the age of 40. Thus, these ages represented the youngest age of a birthing mother and a hypothetical maximum age for births for older women in this community. Compared to the earlier indices, this calculation yields slightly higher levels of reproductive stress for older females and slightly lower estimates for younger and first time mothers, and attributes greater stress to teenaged first-time mothers. The different calculation of reproductive life span was used to lessen a perceived underestimation of stress in older women (Harrington 1983:135) and the potential to overestimate stress among younger or first-time mothers. The reproductive stress index was calculated for women in Yalcoba, but not in Nunoa, because the Nunoa data had gaps in the measurement of lactation and the timing of birth events.

The case studies presented here are based primarily on household health and economic surveys in both communities, and on the reproduction data and women's symptomatology from Yalcoba. The data was collected from a sample of 64 households from one semi-urban town and two rural farming-herding communities in Nunoa, and from 30 households in the village of Yalcoba. Analyses focus on gender differences in illness between adult males and females, and on the way differing economic roles (especially the division of labor) and reproduction stress may affect patterns of illness.

Table 8.1. *Illness symptoms and work lost in adults from a semi-urban town and two rural farming–herding communities in the Peruvian Andes*

| Illness Measure | Semi-urban town | | Rural communities | |
	M	F	M	F
Individuals (N)	28	45	30	40
Percentage symptoms reported*	15.6	22.9	18.8	25.6
Average days lost per adult**	3.2	7.2	3.6	4.8
Annual days lost/household	24.7	62.4	31.2	41.6

* Values averaged across three survey periods.
** Total of six weeks recall (sum of three survey periods – each based on a two-week recall).
M: male; F: female.

Gender differences in health in Nuñoa

Symptomatological surveys in Nuñoa found that the most frequently reported symptoms for all individuals were respiratory (23%), followed closely by musculoskeletal problems, gastro-intestinal illness, headaches and toothaches (Leatherman 1996). The relative order of leading symptoms were the same for men and women. However, adult women reported more illness symptoms and work lost due to illness than men in each of the communities surveyed. Table 8.1 presents the percentage of symptoms reported and the number of work days lost to illness reported in three health surveys, each covering two-week recall periods. Women reported about 7% more symptoms than men in both the farming–herding communities and the semi-urban town. However, gender differences in work days lost were greater in town versus rural households. In both the rural and semi-urban locales, women reported more work disruption due to illness than men, but this was about a one-day difference over six weeks of recall data in rural households versus a four-day difference in the town. If extrapolated to an annual estimate, these data suggest that rural and town men lose 31 and 25 days of work due to illness each year, rural women lose about 42 days, while town women lose over 62 days (Table 8.1). This is an 11-day difference between men and women from rural communities, but a 37-day difference between men and women in the town. Thus, rural–urban differences in the division of labor were not strongly reflected in gender differences in reported symptoms of illness, but may well have been associated with the impact of illness on work.

Work activities of men and women in most of the rural farming–herding households in this part of the Andes were less differentiated than in the town, where household economies were based more on cash incomes from wage work, petty commodity production, and mercantile activities. In the wage labor market, women and men did not have equal access to jobs, and women received half the pay of men for identical jobs (even light agricultural labor). Women in these contexts often contributed less to the household income, and had less direct control and power over household resources and an increased dependency on adult male workers (see Deere 1990:286). Consequently, women tended to work harder and received less for their efforts, and in contexts of seasonally limited food availability, were increasingly vulnerable to problems of undernutrition. Leonard and Thomas (1988), for example, found that women in poorer, semi-urban households experienced much greater seasonal fluctuations in diet, weight, and body fat than either adult males or children. Thus, the relatively higher levels of disruptive illness reported by the town women were perhaps associated with high work loads and marginal diets at a proximate level, and at a deeper level with the social devaluation of their labor and consequent loss of economic power.

If the double-day of women is responsible for greater work days lost in semi-urban women, we should see the greatest work disruption in single female-headed households. Members of female-headed households reported slightly fewer symptoms but more work days lost than in male-headed households. Women in these households reported about 3% fewer symptoms, but lost about 4–5% more work days (Figure 8.1). These differences are not great but add to the gender differences in work days lost in town households. The levels of illness-related work disruption were clearly not related to higher levels of reported symptoms. Rather, slightly lower levels of symptoms were associated with greater effects on work among the women in single-female headed households whose work-days are most constrained.

The specific health problems associated with work days lost to illness in men and women are presented in Figure 8.2. Respiratory problems were associated with most work days lost, which is expected given that respiratory problems lead every health measure for the population: morbidity, mortality, and disease specific clinic data. Unexpectedly, reproductive problems and skeletal muscular problems were ranked second and third in contributing to work days lost in all households (Figure 8.2), and reproductive problems were the leading cause of work days lost to women (40% of lost days). The most common complaint was labeled '*sobreparto*', which was described as illness associated with the stress of childbirth. *Sobreparto* was occasionally used to denote a post-partum infection, but more commonly was used as a folk category to explain a group of symptoms (weakness, general malaise, chronic minor ailments, etc.) which persisted years after the last birth.

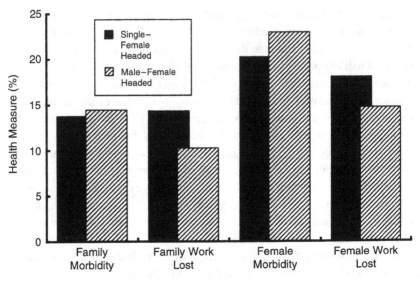

Fig. 8.1 Symptom prevalence and work days lost in single-female headed and male-female headed households in a semi-urban town in the district of Nunoa, Peru.

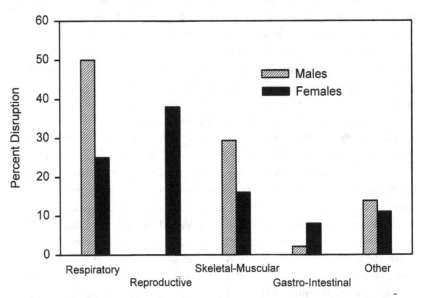

Fig. 8.2 Health problems disrupting work among adult males and females in Nunoa, Peru.

Relationships between high fertility, reproductive stress, and the kinds of illness symptoms described in Nunoa associated with *sobreparto* have been found by other researchers in the Andes and Mexico. Carole Browner (1989) found that Mexican women with four or more pregnancies reported significantly more minor complaints such as headaches, backaches and other body aches, weakness, and fatigue. Katherine Oths (1991) found that women in northern Peru reported symptoms of chronic malaise under the category of *debilidad* which she related to reproductive stress. Ruthbeth Finerman (1983) working with Andean women in Ecuador, associated the prevalence of *nervios* – a folk illness with symptoms related to weakness and nervous feelings – to high fertility levels. Finally, Ann Larme (1993) found that along with the hot–cold properties, reproductive problems were reported by women in Cuyo Cuyo in southern Peru as the leading cause of illness. Similar to my findings in Nunoa, she found that *sobreparto* was an encompassing label for a variety of reproductive illness, and that most *sobrepartos* were associated with similar chronic symptoms of weakness and malaise. As explained by a woman in Cuyo Cuyo, 'After giving birth, a woman's body is completely *malogrado* (ruined), just like after a truck accident' (Larme 1993). Obviously, these women are not talking of specific biomedical disease categories, but rather appear to be highlighting the real and metaphorical stress which they feel comes with repeated childbirth in contexts of heavy work loads and marginal living conditions.

In summary, the information on gender differences in health and women's health status in the context of changing relations of production in Andean households suggests that women may become more vulnerable to illness as gender differences in work roles, economic opportunity, and power, increase. High fertility levels, in conjunction with heavy work loads and marginal diets, might exacerbate vulnerability to illness. The health effects are not always immediate or acute, but can be cumulative and general in nature.

Mayan women in Yalcoba

In earlier research, Magali Daltabuit (1988) reported that women in Yalcoba (Yucatan, Mexico) had heavy work loads and experienced heavy reproductive stress. She suggested that these factors significantly influenced their health. Our subsequent work in 1989 attempted to link these factors to the health status of women and to gender differences in health in the context of male out-migration, which had increased in the previous decades with the growth of a regional tourist economy.

Health surveys were carried out in 30 households in 1989 and 1991, to

Table 8.2. *Household health and gender differences in illness in Yalcoba households, 1989 and 1991 (N=30)*

	1989	1991
Households with illness (%)	63.3	50.0
Sick persons per household	1.0	1.7
Adult females sick (%)	30.0	40.0
Adult males sick (%)	7.0	4.0
Children sick (%)	7.0	9.0

examine economic and gender-based variation in health status. In both years, at least 50% of the households reported an illness in one to two family members in the preceding month, and about half reported illness events serious enough to disrupt their work (Table 8.2). The most prevalent illness categories reported in these health surveys were respiratory, gastro-intestinal, skeletal-muscular, and dermatological. These results were similar to records from a local health clinic (Daltabuit 1988). Households of lower economic status reported approximately 8% more cases of illness, but these differences were not statistically significant. A key difference between moderate and lower income households was that more moderate income families felt that they could obtain help from outside the family in times of illness (25% vs. only 6% of the low income households).

Between 30 and 40% of adult women from the two health surveys reported an illness in the previous month, while only 4–7% of adult males were reported to have been ill. These estimates may be biased by the fact that so many adult males were away during the week, and because a portion received on-the-job health care. Thus, potential health problems might be overlooked or minimized in reports by the adult females who most often provided the information on household health. Nevertheless, the magnitude of differences indicated here suggest a recognizable gender bias in health status. Reasons for these differences might lie in the interaction of heavy work loads, marginal diets, and the biological costs of reproduction (Daltabuit 1988; Browner 1989; Larme 1993). In the present contexts of tourism, increased work loads and the added work and psychosocial stress of being a single head of household for extended periods of the time may exacerbate this interaction. Both men and women reported that women were resilient in the face of illness and kept working through most problems, in part because their efforts were essential to the daily production of the household. Poorer women with little social and

Table 8.3. *Reproductive stress and symptom prevalence among Mayan women from Yalcoba, Mexico*

Symptom category	Low stress (N=12)	High stress (N=18)	Percentage difference low to high stress
All symptoms (%)	27.5	33.3	+5.8
Respiratory (%)	27.5	33.8	+6.3
Gastro-intestinal-urinary (%)	18.6	25.0	+6.4
Heart-circulatory (%)	20.0	22.5	+2.5
Skeletal-muscular (%)	15.0	25.0	+10.0
Nerves (%)	12.0	24.0	+12.0
Weakness (%)	25.0	42.0	+17.0

economic support, in particular, may have had no choice but to work in spite of illness (Daltabuit 1988). Alternatively, some may have had to return to work too soon following birth and post-partum periods which are traditionally viewed as times of weakness when various sanctions against certain work must be followed to prevent illness.

Reproductive histories confirmed earlier findings (Daltabuit 1988) that women in Yalcoba reproduce early and frequently until age 35, and rarely past age 40. Average age at first pregnancy is 19, but 40% of the women were pregnant by age 17, and 12% by age 15. Over their reproductive life, women have between seven and eight pregnancies, one miscarriage, and five to six living children. They breast-feed their babies to 1.5–2 years old, and introduce weaning foods at around six months. The reproductive stress index, a measure of the percentage of a woman's reproductive life spent pregnant and lactating, was calculated for each of the 30 women. Almost one half the women had a medium to low stress index, where less than 60% of their reproductive life was spent pregnant or lactating, and one half had a heavy, or high, stress index of greater than 60%.

Clear health differences were found between these two groups. Women in the high stress category reported 20% more health problems in the previous two-week period, and 25% more ranked their health as poor. They also reported 6% more symptoms than women in the low stress category (Table 8.3). Of primary interest, is that the largest differences between the two groups are seen in symptoms related to skeletal-muscular problems, nerves, and weakness. These are the types of chronic symptom groupings found by Finerman (1983), Browner (1989), and Larme (1993) to be associated with reproduction related illness elsewhere in Mexico and Latin America.

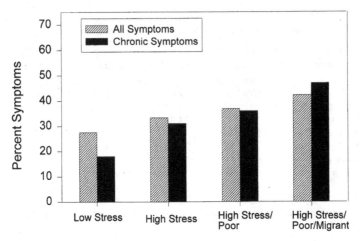

Fig. 8.3 Reproductive stress and symptom prevalence in Yalcoba women, Mexico.

If women's health is due to the combined effects of heavy work loads and high reproductive stress, then high stress women in poor households with spouses who have migrated in search of work should be under the greatest stress. These associations are presented in Figure 8.3. Here, the prevalence of all symptoms and a subset of chronic symptoms are compared in women of low versus high reproductive stress, in women with high reproductive stress from poor households, and in those poor women who have spouses that migrate from the community on a regular basis. Total symptom prevalence increases moderately in these categories (from about 28 to 42%), but chronic symptoms show a more marked increase from about 18% prevalence in low stress women to a 47% prevalence in high stress, poor women with migrant husbands (several of whom are in charge of effectively female-headed households). Small sample sizes limit the analysis, but the data suggest that as work and reproductive loads become increasingly high, health status becomes worse, particularly in terms of chronic fatigue, weakness, pain, and general malaise. In the present contexts of tourism, increased work loads and the added stress of essentially being a single head of household for long durations may exacerbate this interaction.

Summary and discussion

Most contemporary health surveys find consistently higher reports of illness among women than men. Explanations for these differences have focused on women's behavior, knowledge and experiences toward illness, and have suggested for example that: (1) women complain while men are stoic; (2) women's roles as gatekeeper and care giver for the family increase her knowledge of and attentiveness to detecting health problems; and, (3) women are more attuned to their own health status because of their reproductive roles. However, there is more to gender differences in health than reporting behavior. Individuals' experiences and behaviors linked to differing levels of illness are intricately tied to their social and economic position in the household and community. In any given community, variables such as education, access to clean water, material wealth, and occupation are likely to correlate highly with women's health and that of other household members. The point here is that the social relations of production and reproduction in the household, community, and in the regional economy, structure the distribution of these and other correlates to health. They also provide a useful context within which to examine gender differences in health and the processes responsible for increased morbidity and mortality in women found in many health surveys.

Relatively few studies have examined women's health in the household context and relative to other family members. Yet, this is important because processes of household production and reproduction shape the health of all household members, and the reproductive experience and health of women in particular. Taken together, the information from these two populations in Latin America suggests that the multiple roles women play in household production, maintenance, and biological reproduction take their toll in diminished health status. Where the household division of labor was most distinct and men's and women's economic roles and power were most disparate (the semi-urban town in Nunoa, Peru, and the Yucatec Mayan community of Yalcoba), gender differences in illness were greater. Also, where poor economic conditions were associated with marginal diets and heavy work loads, high levels of fertility were associated with higher levels of illness and particularly chronic illness in adult women.

What might these conclusions suggest for studies of prehistoric health? One general implication of this research is to look for gender differences in health in the context of intensification of production. Monetized economies often devalue women's work and economic roles, and in prehistory, more intensified production regimes certainly implied an altered division of labor and potentially an intensification of women's work. Harris and Ross (1987) have noted that agricultural intensification often involves a concomitant pressure on women to increase their productivity and to decrease birth spacing. This in turn puts a burden on women to partition their energy and time between production and biological and social reproduction (Martin 1998). This burden may lead to even more marked gender differences in emerging complex societies. Evidence of skeletal anomalies associated with work stress, arthritis, osteoporosis, trauma, anemia, and infections might all be likely outcomes of changing women's roles in production and reproduction that can be assessed in the skeletal record, and used to evaluate the role of class and gender on health in prehistory.

Rebecca Storey (Chapter 9) suggests that gender differences in skeletal indicators of adult women's health may be a product of lifelong accumulation of stress, and this may well represent a general devaluing of female status that begins in childhood. She suggests that the assessment of gender treatment should begin with an examination of paleopathological indicators that reflect stress during childhood. In her study of the prehistoric site of Copan, Honduras she found no significant gender differences in these indicators of childhood health. This is consistent with modern ethnographic accounts of Mayan communities where no gender specific child neglect has been found. However, in her study, status indicators are significantly associated with indicators of stress in males and females. Status differences are more strongly associated with health stress in males than females. This suggests that the power and resources associated with status might be more accessible to males. It may also indicate that while children of both sexes receive relatively similar treatment and resources, and similar exposure to stressors, adult women may have a proportionately greater exposure to stress due to their structural role in the local economy and added stress from reproduction.

When high fertility is encountered in prehistoric social and economic contexts, it would be surprising if women's health were not strongly affected. For example, Martin (1998) discusses the work of Stodder (1984) on the Mesa Verde population who found more illness in women compared to men in early adult years, and attributed this to childbearing. Adult females also had a higher frequency and more severe manifestation of nutritional anemia. Recently, Martin (1998) has illustrated relationships between gender roles and status, and gender differences in health in different political-economic contexts in a comparison of the La Plata River Valley and Black Mesa in the American Southwest. Little evidence for gender differences in illness was found in the more politically and environmentally marginal Black Mesa area. Conversely, marked gender differences in both illness and violent trauma were found in the La Plata River Valley, which was an environmentally rich, political center, with ample evidence for social status differences in the population. She concludes that by using a political-economic perspective and positioning biological studies within a larger archeological context we can begin to uncover the connections between health and social relations in prehistory.

Conversely, but importantly, an insight that contemporary studies can provide prehistorians is that women's health (and that of other household members) is an important factor in structuring household production and reproduction. Structural inequalities that place women and other household members at risk to illness also often limit their coping options when they become ill (Leatherman 1996). Problems that affect women's health and work will permeate throughout the household limiting basic efforts to maintain the household as a productive and reproductive unit. Thus, studying women's health in the context of household production and reproduction is essential to future work on gender and health in prehistory. Moreover, studying the effects of poor health on household production and reproduction is an added challenge for research on health in prehistory.

References

Beneria L (1979) Reproduction, production and the sexual division of labor. *Cambridge Journal of Economics* 3:203–23.

Beneria L and Roldan M (1987) *The Crossroads of Class and Gender*. Chicago: University of Chicago Press.

Browner C (1989) Women, household and health in Latin America. *Social Science and Medicine* 28:461–73.

Browner C and Leslie J (1995) Women, work, and household health in the context of development. In C Sargent and C Brettell (eds.), *Gender and Health*. Englewood Cliffs: Prentice Hall, pp. 260–77.

Carey JW (1990) Social system effects on local level morbidity and adaptation in the rural Peruvian Andes. *Medical Anthropology Quarterly* 4:266–95.

Daltabuit M (1988) *Mayan Women: Work, Nutrition and Child Care*. PhD dissertation, University of Massachusetts, Amherst.

Daltabuit M and Pi-Sunyer O (1990) Tourism development in Quintana Roo, Mexico. *Cultural Survival Quarterly* 14(1):9–13.

Deere CD (1990) *Household and Class Relations: Peasants and Landlords in Northern Peru*. Berkeley: University of California Press.

de Janvry A (1981) *The Agrarian Question and Reformism in Latin America*. Baltimore: The Johns Hopkins University Press.

Finerman R (1983) Experience and expectation: conflict and change in traditional family health care among the Quichua of Saraguro. *Social Science and Medicine* 17:1291–8.

Goodman A (1998) The biological consequences of inequality in antiquity. In AH Goodman and TL Leatherman (eds.), *Building a New Biocultural Synthesis: Political-Economic Perspectives in Biological Anthropology*. Ann Arbor: University of Michigan Press.

Harrington AJ (1983) Nutritional stress and economic responsibility: a study of Nigerian Women. In M Buvinic, M Lycette, and A McGreevey (eds.), *Women and Poverty in the Third World*. Baltimore: The Johns Hopkins University Press.

Harris M and Ross E (1987) *Death, Sex and Fertility: Population Regulation in*

Preindustrial and Developing Societies. New York: Columbia University Press.

Larme A (1993) Work, reproduction and health in two Andean communities. Working Paper No. 5. In B Winterhalder (ed.), *Production, Storage and Exchange in a Terraced Environment on the Eastern Andean Escarpment.* Chapel Hill: University of North Carolina.

Leatherman TL (1987) *Illness, Work and Social Relations in the Southern Peruvian Highlands.* PhD dissertation, University of Massachusetts, Amherst.

Leatherman TL (1992) Illness as lifestyle change. In R Huss-Ashmore, J Schall, and M Hediger (eds.), *Health and Lifestyle Change.* Philadephia, PA: MASCA, The University Museum of Archaeology and Anthropology, pp. 83–9.

Leatherman TL (1996) A biocultural perspective on health and household economy in southern peru. *Medical Anthropology Quarterly* **10**(4):476–95.

Leatherman TL, Carey J, and Thomas RB (1995) Socioeconomic change and patterns of growth in the Andes. *American Journal of Physical Anthropology* **97**(3):307–22.

Leatherman T and Gordon A (1994) Agrarian transformations and health: introduction. *Human Organization* **53**(4):371–80.

Leonard WR and Thomas RB (1988) Changing dietary patterns in the Peruvian Andes. *Ecology of Food and Nutrition* **21**: 245–63.

Leslie J and Paolisso M (Eds.) (1989) *Women, Work and Child Welfare in the Third World.* Boulder: Westview Press.

Martin D (1998) The American Southwest as a living laboratory: missed opportunities and new directions for the study of human remains. In AH Goodman and TL Leatherman (eds.), *Building a New Biocultural Synthesis: Political-Economic Perspectives in Biological Anthropology.* Ann Arbor: University of Michigan Press. (In press.)

Oths K (1991) *Medical Treatment Choice and Health Outcomes in the Northern Peruvian Andes.* PhD dissertation, Case Western Reserve University.

Painter M (1984) Changing relations of production and rural underdevelopment. *Journal of Anthropological Research* **40**:271–92.

Saitta D (1998) Linking political economy and human biology: lessons from North American archeology. In AH Goodman and TL Leatherman (eds.), *Building a New Biocultural Synthesis: Political-Economic Perspectives in Biological Anthropology.* Ann Arbor: University of Michigan Press. (In press.)

Stodder AW (1984*) Paleoepidemiology of the Mesa Verde Region Anasazi: Demography, Stress, Migration.* MA thesis, University of Colorado, Boulder.

Thomas RB, Leatherman TL, Carey JW, and Haas JD (1988) Consequences and responses to illness among small scale farmers: a research design. In KJ Collins and DF Roberts (eds.), *Capacity for Work in the Tropics.* New York: Cambridge University Press, pp. 249–76.

9

The mothers and daughters of a patrilineal civilization: the health of females among the Late Classic Maya of Copan, Honduras

REBECCA STOREY

Complex patrilineal societies often have strong son preferences. Contemporary demographic and health studies have revealed very biased sex ratios among children in several modern nations. According to Coale and Banister (1994:459–60), 'the ratio of males to females in the population is higher than would be expected from the typical sex ratio at birth and the typical differential mortality. The source of this high male ratio is female mortality that is higher, in relation to male mortality, than would prevail if both sexes had equal access to factors promoting good health.' For instance, in several Asian and African nations, there is strong evidence of daughter neglect, the selective abortion of female fetuses, and perhaps infanticide, as the underlying causes of the bias (see Miller 1981; Hrdy 1990; Coale and Banister 1994). Not surprisingly, along with evidence of son preference in many societies (Hrdy 1990), recent trends in fertility limitation and the ability to prenatally determine sex are exacerbating the bias and creating even higher male:female ratios in countries such as China and the Republic of Korea (Coale and Banister 1994). In these cultures, males are considered more valuable by families and fulfill important societal functions. Consequently, if parents intend to limit their reproduction, the sex of their offspring will be important. In general, in most societies, males are the more valued sex, and females, although important for reproduction and domestic functions, are viewed as less valuable. Critical questions to our understanding of gender patterns of health and disease can be framed using data that indicate that males and females are differentially valued. How, for instance, does the differential valuation of males and females affect the treatment of children? And are 'factors promoting good health' differentially provided according to sex?

Not all nations where son preference might be expected show the biased mortality of young females. For example, males have prominent roles in the families and society in Latin America, and yet these modern nations have no evidence of sex discrimination in mortality (United Nations 1986). The differing patterns of mortality between nations are obviously caused by more than just patrilineal organization or male bias and depend on many other related cultural patterns that affect family formation, child rearing, and care. After all, the production of equal numbers of marriageable males and females is a theoretically strong long-term demographic strategy, especially in cultures practicing monogamy. Biologically, however, males experience higher mortality rates throughout the life span, while females appear to be biologically buffered and survive environmental stresses more successfully (Stinson 1985). In a sense then, boys could benefit from extra care, while some neglect may not seriously harm girls. Many families could practice son preference without necessarily preventing daughters from reaching maturity. Daughter neglect, when not leading to elevated mortality, might be a more pervasive effect of a general worldwide preference for sons and their perceived greater value as adults. Data derived from patrilineal and agnatically-biased societies provide a means to investigate the presence of daughter neglect and its variations in both contemporary and past populations. The presence of a different value and the differential treatment of males and females influences economic, reproductive, political, and personal decisions made by individuals. It is important for our understanding of human health and disease to know how social values placed upon sex may influence or affect health.

One way to explore this issue is to document possible differences in gender treatment in past and contemporary populations. In skeletal populations, it is possible to explore various paleopathological indicators of stress that reflect childhood conditions, but remain visible in the skeleton of adults. With this technique, differences in the presence of childhood stressors in males and females could indicate that one sex might have had to survive more stress in order to reach adulthood, or conversely, one sex may have been more buffered during critical growing years. Skeletal populations are, however, mortality samples, not unbiased reflections of the living populations that produced them (see Wood *et al.* 1992). Thus, individuals dying during childhood are probably different in various ways from those that live to adulthood – perhaps more frail and sick. Hence, they may not reflect variation in child care and childhood environments. Similarly, since skeletally determining the sex of juveniles has yet to be successful, comparing juvenile mortality and illness patterns by sex cannot be done. To test whether there might be differences in the treatment of children by gender, it is necessary to compare those that have survived to maturity, and look for evidence of differences of treatment during their childhood.

Several paleopathological indicators routinely examined in skeletons are caused by stress during childhood growth and remain scorable in adults. These include porotic hyperostosis and cribra orbitalia, the failure to attain full adult stature, and dental defects known as enamel hypoplasias. These lesions and skeletal alterations are the result of chronic conditions, and indicate the presence of an environment where morbidity and malnutrition substantially impacted the lives of individuals, but might not have caused death. Porotic hyperostosis and cribra orbitalia are healed lesions on the cranial vault and orbits, respectively, and are the result of anemia during childhood (see Chapter 4 and Figure 4.1). Their presence have been linked to environments with poor hygiene and high pathogen loads (Stuart-Macadam 1985, 1992). Adult stature is attained through genetic endowment and environmental influence, but short stature is often the result of malnutrition and morbidity (Martorell 1989). Enamel hypoplasias, defects in the enamel as a result of stress that interfered with normal dental formation during growth, also reflect the disease and nutritional status of children in that society (Goodman and Rose 1991; see also Figure 10.4). Individuals who reached adulthood with fewer of these indicators of stress would be expected to have had an easier, healthier childhood than those with more of these chronic indicators of growth arrest and anemia. If one sex proves to have significantly fewer of these stress indicators, it may suggest that as children, one sex received a better diet and a buffered, healthier, or more hygienic environment. Thus, the presence of paleopathological conditions in adult skeletons, in combination with other cultural information, can be used to infer son preference or daughter neglect.

One test of such an approach was recently reported by Lukacs and Joshi (1992). Comparing the prevalence of enamel hypoplasias among modern Rajputs in India, a dominant caste, who display strong son preference and accompanying daughter neglect, the researchers found no statistical differences between the sexes when the presence of enamel hypoplasias was examined. In fact, males displayed slightly higher frequencies. This study, however, focused upon adolescents and young adults, and may not have found sex differences in the presence of enamel hypoplasias because the daughters capable of surviving to these ages were not neglected. The parents probably chose to raise that daughter.

Nevertheless, this attempt brings up a possible complication to this skeletal study – that although severe daughter neglect/strong son preference may be present, it may only be clearly found in children. It is possible that females surviving to adulthood were raised by their families and treated better through childhood than daughters who died as small children. If, however, sons receive preference and better care in a society, but daughters are not neglected to the point of death, then differences in stress levels might be found. This social situation can be tested using the techniques described above.

Background

The Maya of Central America have had a complex society with status differentiation for over 2000 years. One cultural stage of this society, the Classic Maya civilization, flourished in the lowlands of Mexico, Guatemala, Honduras, and El Salvador from about AD 250 to AD 900. Insight into this society, especially into the elites, is provided by a glyphic writing system that has recently been successfully, although not completely, decoded. From the glyphs, the feats and genealogy of rulers have revealed a society where patrilineal descent was practiced, and where warrior prowess was a prerequisite for a successful ruler (see Schele and Freidel 1990). The importance of warfare, the successful capture of other noble males, and the preference for inheritance from fathers to sons among these elite Classic Maya, would lead one to expect the culture to prefer sons and show less interest in raising daughters.

Although there is a definite androcentric bias in the art and glyphs (Haviland 1997), the Classic Maya do portray and mention women. Often a ruler mentions his father and his mother, particularly if she was from another royal family. Women also ruled in some cities, such as in Palenque, and are shown carrying out important blood sacrifices in the ritual cycle at various sites (Schele and Freidel 1990). Females were buried in tombs and display cranial deformation and jade dental inlays similar to those found in elite males, although more males probably received tomb burials and richer offerings at many sites (Haviland 1997). Thus, in spite of patrilineal social organization and a stress on warfare, women, especially elite women, were socially recognized as important. This could be evidence that at least some females would have been treated well throughout their life span.

Unfortunately, there is little evidence about the lives and concerns of the commoners during this period. Evidence from present-day Maya societies, provided by ethnographers and biological anthropologists, gives some insight into how these people might have lived. However, these societies have been affected by European contact and subsequent culture change since the Classic Maya Period (see Chapter 8 for a discussion on the roles that socio-economics play in creating sex differences in disease in some Peruvian and Mexican populations). Today, indigenous Maya populations comprise some of Central and Latin America's most disadvantaged peoples. Research on the modern Maya suggests that the populations are generally patrilineal, with males living and cooperating together. There is also a strong emphasis on the domestic role of women. There is no evidence of great differences in the treatment of boys and girls during childhood, and the recent increasing use of social bilateral ties means wives can bring important kin contacts to the family

(Vogt 1970; Wilk 1991). Thus, strong son preference does not appear to be present here. Sons, though, might often be considered more economically valuable to their families as adults.

Research on health in disadvantaged Maya communities has not revealed evidence of particular daughter neglect when studied by measures of growth (Bogin *et al.* 1992; Crooks 1994). One study of illness and skeletal maturation, however, suggested that there might be some son preference in diet and visits to health providers (May *et al.* 1993).

A recent paleopathological study, using paleopathological indicators of stress similar to the ones chosen here, explored differences in gender treatment during childhood in the colonial Maya site of Tipu (Danforth *et àl.* 1997). Evidence that males were particularly favored during childhood was not found. Similar patterns in frequency of enamel hypoplasias were found between the sexes, but males displayed significantly more porotic hyperostosis. It appears that females may have suffered from fewer chronic stressors due to genetic buffering. No evidence of excessive male preference was found.

The question to be investigated here is whether, in this Classic Maya site, there will be evidence of differential treatment of boys and girls, revealed by the presence of significantly more signs of stress in one sex. Social patterns among the Classic Maya, especially among elite individuals, do not lead researchers to expect strong daughter neglect. Therefore, it is predicted that there will be few differences, if any, seen between males and females. Males, however, might have slightly higher incidences of some conditions, which can be explained by females' stronger biological buffering from environmental stresses (Stinson 1985).

Materials and methods

For the past 20 years, the site of Copan, Honduras, has been the focus of an intensive and large-scale archaeological investigation into the history and characteristics of an important Classic Maya polity. Various excavations have yielded human remains. Most skeletal material is derived from elite households during the time of maximum population and florescence of this society, the Late and Terminal Classic period of circa AD 1000 to AD 700. Two subsamples of this population will be employed in this study. One sample is from a large elite compound in the main residential nucleus of the site. This compound, the House of the Bacabs (Webster 1989), is one of the largest and most impressive of the Late and Terminal Classic period, with 10 court-yards and over 40 dormitories. It housed around 200 people at one time and was home to one of the most powerful noble lineages. The wealth of the

Table 9.1. *Distribution of sexes by status at Copan*

	N	Females		Males	
		n (%)	% of total	n (%)	% of total
Status 1	59	32 (54)	43	27 (46)	46
2	40	25 (62)	33	15 (38)	25
3	35	18 (51)	24	17 (49)	29
Totals	134	75	100	59	100

N: total sample size within each status; *n:* number of individuals in each status; *%:* percentage of individuals out of total number in each status.
Status 1: high status within elite compound;
Status 2: low status within elite compound;
Status 3: rural household sample (outside elite compound).

residents can be recognized by the masonry construction, the presence of carved benches and facade sculpture, and the elaborate tombs. The other sample in this study is from scattered small households of the rural commoners outside the main settlement of Copan. Here, perishable houses were arranged around a simple patio on low mounds and burials generally lacked mortuary goods and formal tombs (Webster and Gonlin 1988). There are some elite compounds in the rural area, but none of those individuals are included here.

The skeletal analysis included only adults whose sex could be confidently assigned. Discriminant functions were derived from individuals who displayed clear morphological indicators of sex on their pelvis and/or cranium. The discriminant functions were then used to determine the sex of more fragmentary individuals on the basis of size and robusticity. Only those individuals who fell clearly (with over 90% probability) into male or female ranges were included in the sample. Thus, the total sample size in this study was 134 individuals, 75 of whom were females and 59 of whom were males. The difference in proportion between females and males was not statistically significant. The elite compound (Status 1 and 2) yielded a total of 99 individuals: 57 females and 42 males. The rural household sample (Status 3) yielded a total of 35 individuals: 18 females and 17 males (Table 9.1).

While status differences between the rural commoners and the elite compound should be expected, the individuals from the House of the Bacabs are probably not all of noble status, judging by differences in mortuary treatments. Because elite males and females might have been better treated, the House of the Bacabs sample was divided into two groups: high status (called Status 1) and low status (called Status 2). Individuals were assigned to Status 1 if they were found in a constructed stone grave and/or had grave offerings. Individuals were assigned to the lower status (Status 2) if there were no grave offerings and burial was in an earthen pit. The lowest status group (Status 3) was comprised of rural commoners. Theoretically, the highest status group (Status 1) will be expected to display the least amount of skeletal indicators of stress, followed by Status 2. Individuals from Status 3, the lowest status group, will be expected to display the greatest amount of stress. Although individual frailty will certainly result in differential survival (Wood *et al.* 1992), the individuals examined in this study had all survived into their adult years. Those with fewer skeletal indicators of stress would be expected to have had an easier, perhaps healthier childhood, than those with chronic, severe skeletal indicators of stress.

Three indicators of childhood stress were examined in this study. Dental enamel hypoplasias, porotic hyperostosis/cribra orbitalia, and adult stature were used as evidence that childhood stress had been survived. The presence of enamel hypoplasias on the maxillary central incisors and the maxillary and mandibular canines was recorded. Only clear macroscopically visible linear pits and grooves were counted. Three categories of presence were created: hypoplasias absent, one hypoplasia present, and two or more hypoplasias present. The presence or absence of porotic hyperostosis/cribra orbitalia was recorded if at least one orbit and a suitable amount of one parietal was available for scoring. Stature was calculated using Genoves (1967) formula derived from Mesoamerican femoral or tibial length. If the femur or tibia was not complete, Steele's segment method (Steele and Bramblett 1988) was used to estimate the complete length of the long bone. Although these are crude estimations of adult stature, they provide indications of relative differences between status groups and the sexes. This comparison is possible because the same calculations, with the same possible errors, are used for all samples.

Statistical tests were performed using SPSS. Since the small sample sizes often will not allow valid chi-square determinations for contingency tables, an unbiased estimate of the true probability of the distribution of the cells was calculated using SPSS Exact Tests (Mehta and Patel 1996). This was accomplished using a Monte Carlo method of repeated sampling of tables of the same dimensions and margin values as the tested contingency table. Statistical significance was set at $p < 0.05$, and accepted only if the 99% confidence interval of the Monte Carlo estimate also was less than 0.05.

The comparison of paleopathological indicators

The demographic profile of the total sample used in this study can be seen in Table 9.1. Status 1 contains more individuals than the other status groups, with Status 1 containing 59 individuals, Status 2 containing 40 individuals, and Status 3 containing 35 individuals. These differences were not statistically significant. Status 2 displays the greatest discrepancy in the number of males and females, with females comprising 62% ($n=25$) of the population within the status group. The differing proportions of females to males were not found to be statistically significant when compared within each status or when females and males from all statuses were simultaneously compared. This finding facilitates comparisons of the frequency of stress, as the numbers of males and females may be treated as equally distributed across each status.

The frequencies of cribra orbitalia and of porotic hyperostosis are provided in Table 9.2. As seen in the table, the majority of individuals in Status 1 and 2 do not display the lesions. The most striking difference appears in Status 3, where most individuals display a lesion. Since Status 3 represents the poorest individuals in the society, it is not surprising that most of these individuals had to survive bouts of anemia to reach adulthood, and that the only individuals with severe lesions are found here. Females from Status 2 display a higher prevalence of both cribra orbitalia and porotic hyperostosis than the males in this group, while male prevalence is slightly greater or equal to females within the other two statuses. For cribra orbitalia or porotic hyperostosis, the differences in the frequency of the lesions between the sexes within a status was not statistically significant. However, when status difference within a sex is explored, a different pattern emerges. For instance, cribra orbitalia and porotic hyperostosis are more frequently encountered in males in Status 1 and in Status 3. The differences in frequency rate between males is statistically significant for both cribra orbitalia ($p=0.02$) and for porotic hyperostosis ($p<0.01$). For females, there was no statistical difference between statuses for cribra orbitalia ($p>0.05$). There was, however, a statistically significant difference for porotic hyperostosis ($p<0.01$). Porotic hyperostosis is frequently found in females within Status 3 and less often found in females within Status 1. This pattern is to be expected when using the model that social status provides access to resources, which in turn, buffers the individual from stressors. Females in the highest social group appear to have been buffered from stressors that provoke iron-deficiency anemia. Thus, while there is no difference between the sexes within a status, status does affect whether a male or a female will have survived conditions associated with anemia during childhood.

Table 9.2. *Frequencies of lesions of cribra orbitalia and porotic hyperostosis*

		N	Absent n (%)	Mild/moderate lesion present n (%)	Severe lesion present n (%)
Cribra orbitalia					
Status 1	Female	16	11 (69)	5 (31)	
	Male	17	10 (59)	7 (41)*	
2	Female	17	9 (53)	8 (47)	
	Male	9	7 (78)	2 (22)*	
3	Female	9	2 (22)	7 (78)	
	Male	8	1 (13)	7 (87)*	
Totals		76	40 (53)	36 (47)	
Porotic hyperostosis					
Status 1	Female	25	22 (88)	3 (12)**	
	Male	18	12 (67)	6 (33)*	
2	Female	22	14 (64)	8 (36)**	
	Male	11	10 (91)	1 (9)*	
3	Female	17	7 (41)	9 (53)**	1 (6)
	Male	13	1 (8)	11 (84)*	1 (8)
Totals		106	66 (62)	38 (36)	2 (2)

N: number of individuals with at least one orbit and suitable amounts of parietal bone; *n:* number of individuals with or without lesion;
%: percentage of individuals out of total number in each status;
*statistically significant ($p < 0.02$) when males compared across statuses;
**statistically significant ($p < 0.01$) when females compared across statuses.
Status 1: high status within elite compound;
Status 2: low status within elite compound;
Status 3: rural household sample (outside elite compound).

Also of interest, is whether there is any pattern among the individuals that have both cribra orbitalia and porotic hyperostosis. Only 38 individuals could be scored for both types of lesions (Table 9.3). Most of these individuals (*n* = 23, 61%) display both lesions, with 11 cases occurring in females and 12 occurring in males. A pattern between the status groups can be found, with Status 1 displaying the fewest cases of both pathological conditions appearing together (*n*=5), Status 2 displaying the next highest frequency (*n*=7), and Status 3 displaying the most (*n*=11). The trend, however, was not found to be statistically significant. Patterns between the sexes within each of the statuses can also be seen. Males within Status 1 more frequently were found

Table 9.3. *Individuals with both porotic hyperostosis and cribra orbitalia*

		N	Both lesions n (%)	Only cribra orbitalia n (%)	Only porotic hyperostosis n (%)
Status 1	Females	6	1 (17)	4 (66)	1 (17)
	Males	8	4 (50)	3 (37)	1 (13)
2	Females	8	6 (75)	2 (25)	0
	Males	1	1 (10)	0	0
3	Females	8	4 (50)	3 (37)	1 (13)
	Males	7	7 (10)	0	0
Totals		38	23 (61)	12 (31)	3 (8)

N: number of individuals with at least one orbit and suitable amounts of parietal bone; *n:* number of individuals with or without lesion;
%: percentage of individuals out of total number in each status;
*statistically significant ($p < 0.02$) when males compared across statuses;
**statistically significant ($p < 0.01$) when females compared across statuses.
Status 1: high status within elite compound;
Status 2: low status within elite compound;
Status 3: rural household sample (outside elite compound).

to display both pathological conditions, while females within Status 2 display both lesions together more frequently than males. Lastly, within Status 3, where the highest incidences of both lesions are found, males display the conditions more frequently than females.

If the occurrence of cribra orbitalia alone is due to a mild or short anemic episode (see Stuart-Macadam 1989), then it can be inferred that Status 1 individuals have been reasonably buffered against stressors that provoke iron deficiency anemia. Within Status 1, nine individuals display only one of the lesions when both could be examined. Within Status 3, however, higher rates of both lesions appearing simultaneously can be found ($n=11$), along with higher rates of the lesions appearing individually ($n=4$). Interestingly, similar to the pattern found when only one lesion type was examined (Table 9.2), Status 2 males and Status 1 females appear to display particularly low frequency rates, suggesting that these groups might have been buffered. Caution must be exercised, however, since the sample sizes are so small.

The frequency of enamel hypoplasias is provided in Table 9.4. Out of the original sample size of 134 individuals, 117 (87%) had canines that could be scored, and 83 (62%) had scorable central incisors. Examining individuals with canines, only 9 (8%) displayed no sign of enamel hypoplasia. For individuals with incisors, 19 (23%) were free from the lesion. The overall patterns

Table 9.4. *Frequencies of enamel hypoplasias by sex and status at Copan*

	N	No enamel hypoplasia n (%)	One or more enamel hypoplasia n (%)
Permanent canines			
All Females	66	5 (8)	61 (92)
All Males	51	4 (8)	47 (92)
Totals	117	9 (8)	108 (92)
Permanent incisors			
All Females	46	11 (24)	35 (76)
All Males	37	8 (22)	29 (78)
Totals	83	19 (23)	64 (77)

		N	No enamel hypoplasia n (%)	One enamel hypoplasia n (%)	Two or more enamel hypoplasia n (%)
Permanent canines					
Status 1	Female	29	2 (7)	14 (48)	13 (45)
	Male	24	1 (4)	12 (50)	11 (46)
2	Female	22		14 (64)	8 (36)
	Male	13		5 (38)	8 (62)
3	Female	15	3 (20)	8 (53)	4 (27)
	Male	14	3 (21)	8 (57)	3 (21)
Permanent incisors					
Status 1	Female	23	4 (17)	8 (35)	11 (48)
	Male	19	4 (21)	8 (42)	7 (37)
2	Female	15	2 (13)	5 (33)	8 (53)
	Male	12	2 (16)	5 (42)	5 (42)
3	Female	8	5 (63)	2 (25)	1 (13)
	Male	6	2 (33)	2 (33)	2 (33)

N: number of individuals with a scorable tooth; n: number of individuals with or without lesion; %: percentage of individuals out of total number of each sex.
Status 1: high status within elite compound;
Status 2: low status within elite compound;
Status 3: rural household sample (outside elite compound).

of lesion frequency are similar between the canines and central incisors, as both tooth types indicate that huge proportions of males and females displayed enamel hypoplasias. For instance, 61 (92%) females with canines, and 35 (76%) females with incisors display the lesion, while 47 (92%) males with canines, and 29 (78%) males with incisors display the lesion. There are

Table 9.5. *Stature estimation for the Copan skeletons*

	Sex	N	Estimation	SD
Status 1	Females	25	155.2 cm[a]	±4.3
	Males	19	163.5 cm[a,b]	±3.5
2	Females	16	155.4 cm[a]	±2.2
	Males	9	162.9 cm[a,b]	±2.9
3	Females	13	154.9 cm[a]	±3.4
	Males	9	160.1 cm[a,b]	±2.9

[a]Statistically significant differences ($p < 0.01$) between male and
female heights within each status using T-tests.
[b]Statistically significant differences between males in each status (F=
3.44; $p = 0.04$) using oneway ANOVA.
Status 1: high status within elite compound;
Status 2: low status within elite compound;
Status 3: rural household sample (outside elite compound).

no statistically significant differences in the presence of enamel hypoplasias
occurring in canines or incisors between the sexes. Patterns of frequency
between the sexes within each status were also similar, except within Status
2, where 64% ($n = 14$) of the females with canines display only one hypoplasia.
Of the males with canines in Status 2, 62% ($n = 8$) display two or more
lesions. While it generally appears as though more individuals, both males
and females, have two or more hypoplasias in Status 1 and 2, the differences
are not statistically significant. Thus, unlike the case with porotic hyperostosis/
cribra orbitalia, status does not appear to be correlated with the presence of
this stress indicator. In general, childhood stressors provoking growth arrest
in enamel formation appear to be widespread in the Copan sample, irrespective
of status.

The stature estimates for each sex within each status group are provided
in Table 9.5. Females are significantly shorter than males in all three statuses,
with mean heights generally declining by status. The mean height of females
between the status groups were not statistically significant. Between males
from each status group a statistically significant difference was found. Status
1 males are substantially taller than Status 3 males. This finding is similar to
that found by Haviland (1967) for the Late Classic at the site of Tikal, where
males in tombs were taller than other males. This might be evidence that
males, especially elite males, may be more advantaged. If the mean female
height is compared to the mean male height within each status, Status 1 and
2 females reach 95% of the male height, and within Status 3, females reach
97% of male height. Although accounting for natural fluctuations in sexual

dimorphism when sample sizes are low is extremely difficult, it could cautiously be postulated that females from the House of the Bacabs are stunted in relation to the males. This would imply that females were not provided the resources needed to achieve their genetic potential for stature. It is more probable, however, that Status 3 males are stunted, bringing them closer to the mean height of females. Male stature appears to be more sensitive to environmental conditions in many populations around the world. Thus the greater difference in stature between males and females in the higher status groups may be due to nutritional and disease buffering among the elite. In a recent article by Marquez and del Angel (1997), stature estimates of precolumbian Mayas in Mexico during the Classic period suggest that females averaged 148 cm, and males 158 cm, rendering females 94% of the height obtained by males. Therefore, it might be that Copan females are generally more advantaged than many Maya populations. In the present study, stature does not indicate any particular disadvantage in the childhood environment of females, who seem to be reaching a similar adult height whatever their social status in the society. Males, however, do appear to be either privileged or stunted by their status.

Conclusion

The three paleopathological indicators studied here, porotic hyperostosis/ cribra orbitalia, dental enamel hypoplasia, and adult stature, do not reveal any differences in frequency of childhood stress markers between males and females of the same socio-economic status. There are statistically significant differences in the frequency of lesions associated with childhood anemia and male adult stature. These differences are associated with status differences, indicating that these markers can be informative about differences in childhood environment and health. The elite individuals tend to have fewer incidences of anemia and achieve a taller stature than the others, an indication that they were successfully buffered from the effects of malnutrition and disease. All status groups displayed relatively high frequencies of enamel hypoplasias, indicating that stress sufficient to cause enamel growth disruption was ubiquitous in the Copan society of the Late Classic. There was a trend for elite individuals to have two or more hypoplasias more often than the rural commoners. Although the difference only approached significance, it could be that elite individuals were better able to survive multiple stress episodes than commoners. Individuals within Status 3 with more than one such stress episode may have died as children. This could be another indication of the better childhood environment for elite individuals in this society.

The stress markers chosen for evaluation in this study did not provide any evidence of son preference or daughter neglect among the Late Classic Maya of Copan. It could be that the techniques chosen were not sensitive enough to indicate differential treatment of children, or that, as discussed before, adult females represent those children invested in by their parents, while neglected daughters died during childhood. The patrilineal inheritance of Maya society, and the importance of warfare during the Late Classic, might predispose this culture to value males and devalue females. Males are certainly more prominent in the iconography, and are more likely to have a tomb burial. But, females can play an important role in the elite iconography, and can be buried in tombs with males and by themselves. In fact, there is evidence that elite females were valued as wives and mothers, and such attitudes could have pervaded Late Classic Maya society in general.

The paleopathological conditions examined in this study provide information that is not indicative of gender discrimination, but might affect the health of children by predisposing females to more stress or providing males with adequate buffering mechanisms. The pathological conditions do suggest possible discrimination by socio-economic status, with rural commoners having to survive more stress as children to reach adulthood. However, regardless of status, stressors were present and powerful in the lives of children.

The Late Classic Maya society was the apogee of Copan. It was a society that was to 'collapse' and disappear shortly after AD 1000 in many parts of the Central American lowlands (Culbert 1988). The frequent evidence of stress associated with anemia, growth arrest in dental enamel, and the evidence of decreased stature in some males, indicates a general health and nutritional environment for children that was less than optimal. This is a society apparently suffering from social structural problems and probable food shortages. These conditions might eventually have caused its end. In such situations, it would not be surprising if individuals and families would have made decisions to favor some children for various reasons that probably varied by circumstance and status. However, for those that lived to their adult years in the patrilineal Late Classic Copan culture, there is little evidence that males were preferred and females neglected during childhood.

Further investigation into the issues of daughter neglect or son preference is crucial. It is essential that we continue to explore patterns of the past, using different skeletal indicators in adult and juvenile skeletons. Research into contemporary societies is also needed in order to gain perspective on the prevalence of gender-based practices and their effects upon culture. It is possible that contemporary gender patterns are recent phenomena, caused by particular economic and social pressures of the late twentieth century. This notion, through the careful use of paleodemographic and paleopathological tools, can be tested. Finding evidence of the presence of gender preferences in the past can begin to assist us in understanding modern demographic patterns and cultural pressures.

Acknowledgments

The Copan osteological study has been conducted with the permission and support of the Instituto Hondureno de Antropologia e Historia. I have also received support from the Fulbright Program and the University of Houston. I gratefully acknowledge the help and feedback that I have received from students and colleagues through the years; this study could not have been done without the help of many individuals.

References

Bogin B, Wall M, and MacVean RB (1992) Longitudinal analysis of adolescent growth of *Ladino* and Mayan school children in Guatemala: effects of environment and sex. *American Journal of Physical Anthropology* **89**:447–58.

Coale AJ and Banister J (1994) Five decades of missing females in China. *Demography* **31**:459–79.

Crooks DL (1994) Growth status of school-age Mayan children in Belize, Central America. *American Journal of Physical Anthropology* **93**:217–28.

Culbert TP (1988) The collapse of Classic Maya Civilization. In N Yoffee and GL Cowgill (eds.), *The Collapse of Ancient States and Civilizations*. Tucson: University of Arizona Press, pp. 69–101.

Danforth ME, Jacobi KP, and Cohen MN (1997) Gender and health among the colonial Maya of Tipu, Belize. *Ancient Mesoamerica* **8**:13–22.

Genoves S (1967) Proportionality of the long bones and their relation to status among Mesoamericans. *American Journal of Physical Anthropology* **26**:67–77.

Goodman AH and Rose JC (1991) Dental enamel hypoplasias as indicators of nutritional status. In M Kelly and C Larsen (eds.), *Advances in Dental Anthropology*. New York: Wiley-Liss, pp. 279–93.

Haviland WA (1967) Stature at Tikal, Guatemala: implications for ancient Maya demography and social organization. *American Antiquity* **32**:316–25.

Haviland WA (1997) The rise and fall of sexual inequality: death and gender at Tikal, Guatemala. *Ancient Mesoamerica* **8**:1–12.

Hrdy SB (1990) Sex bias in nature and in history: a late 1980s reexamination of the 'biological origins' argument. *Yearbook of Physical Anthropology* **33**:25–37.

Lukacs JR and Joshi MR (1992) Enamel hypoplasia prevalence in three ethnic groups of Northwest India: a test of daughter neglect and a framework for the past. In AH Goodman and L Capasso (eds.), *Recent Contributions to the Study of Enamel Developmental Defects. Journal of Paleopathology*, Monographic Publication **2**:359–71.

Martorell R (1989) Body size, adaptation, and function. *Human Organization* **48**:15–20.

Marquez L and del Angel A (1997) Height among prehispanic Maya of the Yucatan Peninsula: a reconsideration. In SL Whittington and DM Reed (eds.), *Bones of the Maya: Studies of Ancient Skeletons*. Washington, DC: Smithsonian Institution Press, pp. 51–61.

May RL, Goodman AH, and Meindl RS (1993) Response of bone and enamel formation to nutritional supplementation and morbidity among malnourished Guatemalan children. *American Journal of Physical Anthropology* **92**:37–52.

Mehta CR and Patel NR (1996) *SPSS Exact Tests 7.0 for Windows*. Chicago: SPSS, Inc.

Miller BD (1981) *The Endangered Sex: Neglect of Female Children in Rural North India*. Ithaca: Cornell University Press.

Schele L and Freidel DA (1990) *A Forest of Kings*. New York: William Morrow and Company, Inc.

Steele DG and Bramblett CA (1988) *The Anatomy and Biology of the Human Skeleton*. College Station: Texas A&M University Press.

Stinson S (1985) Sex differences in environmental sensitivity during growth. *Yearbook of Physical Anthropology* **28**:123–47.

Stuart-Macadam P (1985) Porotic hyperostosis: representative of a childhood condition. *American Journal of Physical Anthropology* **66**:391–8.

Stuart-Macadam P (1989) Porotic hyperostosis: relationship between orbital and vault lesions. *American Journal of Physical Anthropology* **80**:187–93.

Stuart-Macadam P (1992) Anemia in past human populations. In P Stuart-Macadam and S Kent (eds.), *Diet, Demography, and Disease: Changing Perspectives on Anemia*. New York: Aldine de Gruyter, pp. 151–70.

United Nations (1986) *Age Structure of Mortality in Developing Nations*. New York: United Nations.

Vogt EZ (1970) *The Zinacantecos of Mexico: A Modern Maya Way of Life*. New York: Holt, Rinehart, and Winston.

Webster DL (Ed.) (1989) *The House of the Bacabs. Studies in Pre-Columbian Art and Archaeology* No. 29. Washington, DC: Dumbarton Oaks.

Webster DL and Gonlin N (1988) Household remains of the humblest Maya. *Journal of Field Archeaology* **15**:169–90.

Wilk RR (1991) *Household Ecology: Economic Change and Domestic Life Among the Kekchi Maya in Belize*. Tucson: University of Arizona Press.

Wood JW, Milner GR, Harpending HC, and Weiss KM (1992) The osteological paradox: problems of inferring prehistoric health from skeletal samples. *Current Anthropology* **33**:343–70.

10

A history of their own: patterns of death in a nineteenth-century poorhouse

ANNE L. GRAUER, ELIZABETH M. McNAMARA, and DIANE V. HOUDEK

The complex interplay between human culture and biology is an important subject of anthropological research. While cultural anthropologists, such as Fox (1980) and Martin (1987), note that culture can be shaped by human biology, biological anthropologists such as Armelagos and McArdle (1975), Buikstra (1981), and Cohen and Armelagos (1984) (to name but a few), note that human biology can itself be shaped by culture. Within the field of paleopathology in particular, understanding the relationships between human culture and patterns of health, disease, and death is becoming increasingly important.

The adoption of this 'biocultural' approach in paleopathology provides exciting avenues of research. Studies of human patterns of morbidity and mortality can be framed within a larger context of biotic and abiotic environments. The analysis of human skeletal remains can go beyond exploration of patterns of life and death. Social conditions within which a population existed can be elucidated. In this chapter we offer an example of how paleopathological research can provide insight into different roles and opportunities afforded to individuals based upon their sex. Such differences, we assert, exist today and impact disease in twentieth-century populations.

The specific focus of this chapter is on life and death in a nineteenth-century poorhouse. Poverty in the United States, according to many social historians, was tightly linked to economic opportunities, immigration patterns, developing urbanization, and social structures. The last created different environments for economic and social stability depending on one's sex. Likewise, one's sex, and consequently his or her gender role, could also lead to economic and social disenfranchisement. Stansell (1986:56) asserts that 'the working poor, especially immigrant women, were able to create urban communities in the context of massive transiency.' Traditions of cooperation and mutual aid provided support systems amongst women (Weatherford 1986; Harzig 1997). However, women were the most economically vulnerable. 'Poor men were able to support themselves until relatively late in their lives and avoided the poorhouse longer than women. Although women more often succeeded in avoiding it altogether, when they did enter they did so younger than men because of their vulnerability and lack of employment opportunity' (Katz 1986:87). The goal of this chapter is to explore whether paleopathological analysis can assist in revealing the presence of gender differences in social conditions. As noted by Katz (1986:58), 'even poorhouses had a history of their own.'

The sample

A skeletal sample of 120 individuals, dating 1851–1869, was exhumed in 1990 on the west side of Chicago. The skeletons were from a cemetery associated with the Cook County Poor Farm, established in 1851, and located at that site until 1869 (Keene 1989, 1991; Grauer 1992; Grauer and McNamara 1993, 1995; Grauer *et al.* 1995). According to the *Daily Democratic Press* (December 14, 1854) the almshouse (known also as the Dunning Poorhouse) included a three-storey brick building, with one two-storey wing reserved for the care of the county insane (Brown 1941. *Daily Democratic Press*, Dec.14, 1854). Medical care was provided by the city physician. Also allegedly provided were regular meals, proper heating, separate housing for men, women, and children, separate facilities for the insane, and useful employment (Houdek 1995:12). A report, however, filed in 1870 by the Board of State Commissioners of Public Charities of the State of Illinois sharply criticized the facility for miserable planning, bad management, and overcrowding. Other documentary information concerning the operation and occupation of the facilities is sparse (probably due to the Chicago Fire of 1871). Consequently, the paleodemographic and paleopathological analysis of this cemetery population was utilized to provide insight into the social environment of these otherwise 'invisible' people.

Methods and techniques

Skeletal age at death and sex were assessed using multi-variate nonmetric techniques described in *Standards for Data Collection from Human Skeletal Remains* (Buikstra and Ubelaker 1994). These techniques included macroscopic evaluation of dental eruption and formation rates, and the degree of epiphyseal formation and fusion in subadult material. Age at death for adult material was assessed through the examination of the pubic symphysis and auricular surface morphology, dental attrition rates, and degree of ectocranial suture closure. Sex was determined using morphological attributes of the os coxae and cranium. A maximum number of techniques was employed independently for each skeleton. Age and sex were determined by three or more independent groups of researchers. The final age and sex designations were achieved by averaging group conclusions. In total, 105 skeletons were assigned to an age-at-death interval. Only individuals determined to have died over the age of 20 years were classed as 'adults' and included in this study. Subsequently, 52 individuals could be assigned to both a sex and an age at death category.

Our paleopathological evaluation focused upon the presence of general indicators of stress in bone and dental material. The presence of porotic hyperostosis, often synonymously referred to as cribra orbitalia when found in the orbits of the frontal bone, is characterized by a spongy or porous appearance of the crania, caused by diploic thickening and thinning of the outer table of the bone. It is clinically and archaeologically associated with the occurrence of childhood anemia (Stuart-Macadam 1985; and Chapter 4), and was recorded when present on the orbital roofs and calvaria (Figure 10.1). Periosteal reaction, also referred to as periostitis, is a skeletal lesion associated with the presence of nonspecific systemic infection. The lesion was recorded on anatomical features displaying areas of localized or non-localized prolifer-

Fig. 10.1 Porotic hyperostosis displayed an adult cranium.

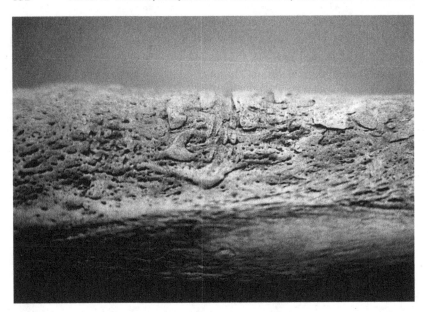

Fig. 10.2 Periosteal reaction displayed on an adult tibia.

ative hypervascular bone. The lesions can have a scab-like appearance, or can alter the cortex and medullary cavity of the bone, producing the appearance of 'swelling' (Figure 10.2). Trauma, identified by characteristic alterations of the shape, size, and contours of bones was also recorded (Figure 10.3). The presence of linear enamel hypoplasia (LEH), representing areas of abnormal enamel formation caused by various systemic stressors occurring during childhood development (Goodman *et al.* 1993) was recorded when detected macroscopically (Figure 10.4). Only LEH occurring on the canines was included in this study. Dental health was evaluated by recording the presence of caries (dental decay commonly referred to as cavities) on the molars. It was also evaluated by the presence of periodontitis (gum disease), which results from the inflammation of the gingival tissue surrounding the teeth (Figure 10.5). Both of these conditions can provide insight into diet and dental hygiene.

Results and discussion

Demographic profile

Of the 61 adults in the Dunning sample assigned a sex, 56% ($n=34$) are female and 44% ($n=27$) are male (Table 10.1). The difference is not statistically significant. The pattern of mortality by age and sex is also strikingly

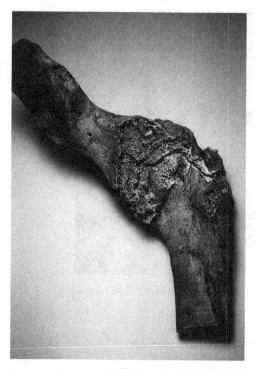

Fig. 10.3 Trauma of the hip joint causing complete fusion of the femur to the acetabulum.

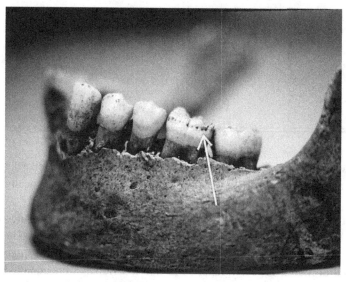

Fig. 10.4 Linear enamel hypoplasia (arrowed) of the mandibular M1.

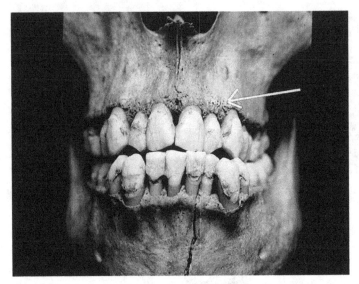

Fig. 10.5 Periodontitis of the maxillary and mandibular alveoli (arrowed).

Table 10.1. *The demographic distribution of adults from the Dunning population*

Age	Females	Males	Sex undetermined
20–24	7	8	1
25–34	8	8	1
35–44	7	6	0
45–54	4	2	0
55+	1	1	0
Total	27	25	2
Undetermined Age	7	2	5
Total (*n*:68)	34	27	7

similar between the sexes. A Kolmogorov–Smirnov test indicated that the cumulative age difference between females and males was not statistically significant.

Historical sources, such as *The History of Public Assistance in Chicago 1833–1893* (Brown 1941), concur that most residents of Chicago's first poorhouse were between 25 and 35 years of age (36% in 1848). *The Chicago Daily Journal* of March 11, 1858, reports that 435 individuals were admitted to the Dunning facility in the winter of 1857–1858. Of those admitted, 40%

were female and 60% were male. A second report, issued September 15, 1858, states that 350 individuals were admitted during the summer months, most were between 20 and 30 years of age, with 30- and 40-year-olds comprising the second largest age group. Of those admitted 46% were female, 54% were male. Similarly, historical records from 1857 and 1858 indicate that 91% of the Dunning Poorhouse residents were immigrants, most were from Ireland, and most were male.

Thus, while historical sources portray a poorhouse population similar to our cemetery population in terms of age, they indicate that more males than females were admitted. This male predominance is not reflected in the cemetery sample. Several hypotheses can be offered in explanation. First, there is a possibility that our method of aging and sexing was flawed. Weiss (1972) and Walker (1995) have noted that many archaeological populations contain more males than females, whereas sex ratios in living populations are the opposite. Walker (1995) asserts that 'sexism in sexing' is at work, leading paleopathologists to rely upon cultural stereotypes of 'typical' female or male cranial morphology. Walker warns that young males are often classified as females, and that skulls of elderly women are often misclassified as males. We argue, however, that this bias was circumvented by basing sex determination on morphological variation in the os coxae, which does not rely upon the detection of 'gracile' or 'robust' features. Our population includes a greater number of females than males, contrary to Walker's expectations. Almost equal numbers of females ($n=15$) and males ($n=16$) between the ages of 20 and 34 years of age appear in our sample, while the number of females determined to be over the age of 35 years surpasses that of males ($n=12$ females, 9 males).

Another explanation for the greater number of females in the cemetery sample might be biased burial practices, with fewer men dying and being buried at the facility. Being buried elsewhere, however, took financial resources, of which residents of the poorhouse had few. Contrary to the possible bias towards females in the cemetery sample, Katz (1986:87) suggests that women without resources were more likely to be taken care of by their children. Upon death, therefore, children may have been more inclined to pay for the burial of their mother in a cemetery other than one at a poorhouse. Furthermore, admission to the poorhouse was usually a last resort, being used when social ties were lacking. If poorhouse residents had no social ties to serve as a safety net in life, in death they were also likely to be abandoned. Thus, we assert that our cemetery sample is unlikely to have been influenced by the biases applicable to some cemetery populations.

Consequently, our contention is that the demographic patterns of the Dunning Cemetery population can be accepted as a valid indication of patterns of *death* at the poorhouse. Our data indicate that although more males entered the facility, a higher proportion of entering females never left.

Paleopathological assessment

Results of the examination for pathological conditions present on the skeletal remains from the Dunning Cemetery appear in Figure 10.6. As indicated, porotic hyperostosis was present on 26% ($n=6/23$) of adult females with crania available for examination, and on 42% ($n=11/26$) of adult males with crania. Periosteal reaction was noted on 26% ($n=7/27$) females in the sample, and on 24% ($n=6/25$) of the males. Trauma was recorded on 22% ($n=6/27$) of the females and on 24% ($n=6/25$) of the males. Linear enamel hypoplasia was found on the canines of 32% ($n=8/25$) of the females, and on 42% ($n=10/24$) of the males. The assessment of dental health, measured by the presence of carious molars and periodontitis, indicated that 44% ($n=11/25$) of the females and 56% ($n=14/25$) of the males had a least one carious molar. Periodontitis was found along the alveolar margins of either the maxilla or mandible in 73% ($n=19/26$) of the females and 76% ($n=19/25$) of the males. The differences in frequency of the conditions in females and males were not statistically significant.

At first glance it appears that disease and trauma played a similar part in the lives of the women and men in the Dunning Cemetery sample. Females, so it seems, were as likely to have suffered through childhood bouts of anemia

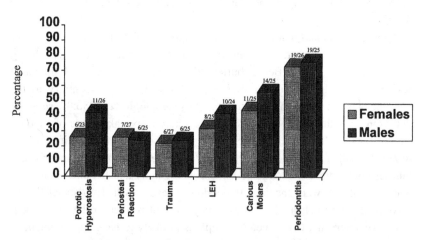

Fig. 10.6 The frequency of pathological conditions in females and males from the Dunning Cemetery population. LEH: linear enamel hypoplasia.

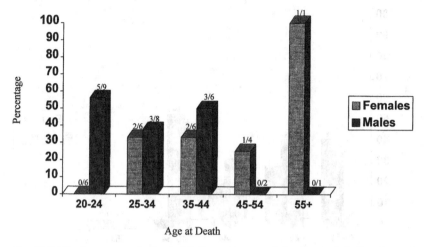

Fig. 10.7 The frequency of cribra orbitalia and/or porotic hyperostosis in females and males from the Dunning Cemetery population.

(indicated by the presence of porotic hyperostosis) and severe childhood stressors (indicated by the presence of linear enamel hypoplasia) as men. Overcoming infection (inferred from the presence of remodeled periosteal reaction) was common in both sexes. Both females and males suffered similar rates of trauma and both endured poor dental health.

However, a different picture emerges when demographic patterns of the pathologies are explored. Recognizable differences in the occurrences of porotic hyperostosis, periosteal reaction, LEH, periodontitis and caries by age and sex are present. For instance, the frequency of porotic hyperostosis (Figure 10.7) is greater in males who died between the ages of 20 and 44 years, than in females in any age category. This is especially apparent in 20- to 24-year-olds, where none of the six females in this age group displays the lesions, compared to five (56%) out of nine males. This pattern is repeated in 20- to 24-year-olds for the occurrence of periosteal reaction (Figure 10.8). Here, no female displays the lesion. In age groups over 24 years, however, a higher proportion of females display the condition. The demographic pattern of LEH (Figure 10.9) indicates that a greater percentage of males (57%, $n=4/7$) within the 20- to 24-year-old age group display this dental deformity compared to females (43%, $n=3/7$). No female is recorded with LEH in the 25- to 34-year-old age group. The pattern of periodontitis is similar (Figure 10.10), with 100% ($n=8/8$) of males between 20 and 24 years of age displaying the condition, while the percentage of females (29%, $n=2/7$) in the age group with the condition is substantially less. Again, the percentage of females with

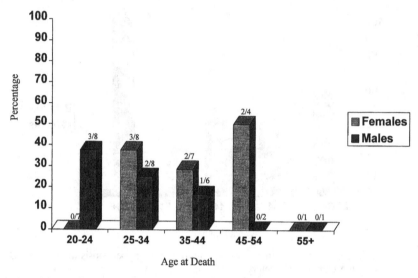

Fig. 10.8 The frequency of periosteal reaction in females and males from the Dunning Cemetery population.

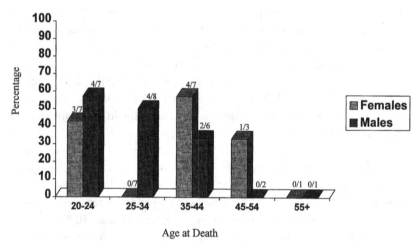

Fig. 10.9 The frequency of linear enamel hypoplasia in females and males from the Dunning Cemetery population.

the pathology is greater in age groups above 24 years. The occurrence of carious molars also varies according to age and sex (Figure 10.11), with 63% (n=5/8) of all males 20–24 years old with molars displaying the condition. Only 14% (n=1/7) of females in that age group have carious molars. As with the other pathologies discussed above, the pattern changes for older adults.

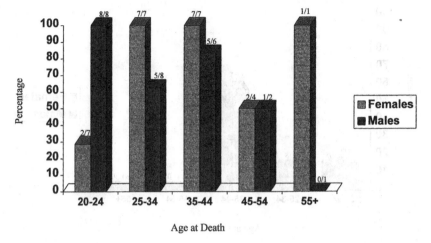

Fig. 10.10 The frequency of periodontitis in females and males from the Dunning Cemetery population.

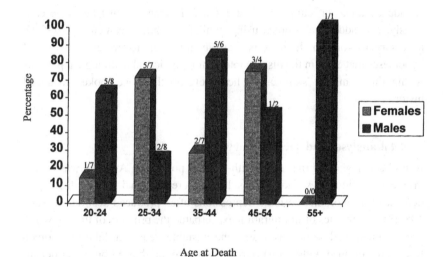

Fig. 10.11 The frequency of carious molars in females and males from the Dunning Cemetery population.

The pathological condition with the greatest degree of demographic similarity is trauma (Figure 10.12). Unexplainably, trauma occurs in equally low percentages for females and males in all age categories except the 35- to 44-year-old age group, where it rises substantially for both sexes.

While demographic patterns of disease and trauma in the Dunning sample

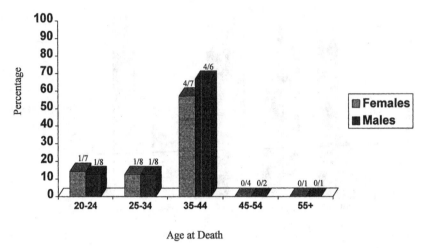

Age at Death

Fig. 10.12 The frequency of trauma in females and males from the Dunning Cemetery population.

are identifiable, our sample sizes are small. Inferences must be made cautiously, and additional studies using similar poorhouse cemeteries, will be necessary to confirm them. It is feasible, however, to present conditional hypotheses that explain the discernable demographic and pathological patterns within this cemetery sample. To help develop these we looked to social history.

Skeletal analyses and social history

It has been suggested that nineteenth-century poverty in America was rooted in the condition of working-class life, where seasonal work, fluctuating demands for labor, and periodic depressions often produced destitution (Katz 1986:89). High rates of immigration exacerbated the problem of finding work. 'For women, paid work was sparse and unstable, requiring laboring women to remain confined within a patriarchal economy predicated on direct dependence on men' (Stansell 1986:18). The result, according to Katz (1983), was that most women, upon entering a poorhouse, listed no occupation. Those that did, usually had been domestic servants. Because most had not worked for wages, women became dependent on welfare earlier than men, especially if they were widowed. Young women used the almshouse as a temporary refuge when they could not rely on the income of a man, when they were unable to find work, or when they were ill. Some older women without family support and unable to work would spend their final years in the facility (Stansell 1986).

Demographic and paleopathological analyses of the Dunning Cemetery sample confirms that nineteenth-century social conditions and economic opportunities were varied and complex. The greater number of females in the Dunning Cemetery sample suggests that in spite of the greater number of men admitted to the institution, a greater percentage of the women residents died there. This comes as no surprise, as historical records imply that young men were more often short-term residents of poorhouses, using the facility to tide them over when work was hard to find (Katz 1986).

If the social historians are correct in asserting that women entered the poorhouse at younger ages than men, then the patterns of morbidity within the 20- to 24-year-old age group at Dunning are particularly interesting. For instance, while the number of females and males are nearly equal in this age group (7 females, 8 males), a consistently smaller proportion of young females display the pathological conditions we examined. Young women, it appears, were not entering and dying in the poorhouse with a legacy of childhood anemia, bouts of infection, evidence of enduring severe non-specific stress, and poor dental health. More likely, they were dying of acute conditions contracted shortly before their deaths. Their presence in the cemetery sample suggests that poor health and difficult childhoods did not bring them to the facility, and that regardless of reasonable health upon entering, their prolonged (if not eternal) residency put their lives in jeopardy. A greater proportion of older females buried at the Dunning Cemetery display pathologies. These individuals might represent the widows and the infirm who, upon entering the facility, could expect to die there. Poorhouses harbored notoriously insalubrious environments.

Young males buried in the poorhouse cemetery present a different picture. The high proportion of 20- to 24-year-old males displaying skeletal pathologies suggests the presence of three disadvantages: (1) They had withstood stressful childhoods (apparent by the percentage displaying porotic hyperostosis and LEH); (2) they had suffered from poor health and dental disease before entering the facility (apparent by the percentage displaying periosteal reaction and periodontitis); and (3) they may have had few, if any, options but to stay. According to Stansell (1986:7), 'the almshouse was in part populated by men of working age who suffered from running sores and broken limbs.' Healthy young men, while constituting the majority of individuals admitted into the poorhouse, also comprised the greatest number of those discharged.

Conclusion

The paleopathological record compliments the historical record in showing that the lives of poor women and poor men differed considerably in nineteenth-century America. Young females who died at the Dunning Poorhouse appear to have entered in a healthier condition than men of the same age. Upon entering, however, they stayed longer and more often died there. Lack of economic opportunities, reliance on men for economic stability, and severed social ties were likely causes of female destitution. This situation often led healthy women, young and older, into lives completely dependent upon institutionalized care. Eventually, with women less likely to find a way out of the poorhouse, the harsh environment took its toll. Thus, in spite of fewer numbers of women being admitted to the poorhouse, a greater proportion of them would become a part of the cemetery population. Young males buried in the cemetery appear to have experienced a different history. They were survivors of stressful childhoods, and were probably unable to secure jobs due to illness or disability. A high proportion of them were immigrants, making them unlikely to have wide social networks. These young men, while making up a small proportion of men who were admitted to the poorhouse, were the most likely never to leave.

In summary, we assert that many biocultural factors played a role in the morbidity and mortality of individuals in this cemetery population. The paleopathological analysis provided insight into the biology of the population. Information provided by social historians, sociologists, cultural anthropologists, and archivists contributed to the development of our biocultural approach. The interpretation of the skeletal data, for instance, required a consideration of the influences of cultural biases, including those relating to historical burial practices, and to those incorporated into methods currently used to make sex and age determinations in skeletal material. In this study, paleopathological data complemented the recorded data. The creation of clear distinctions between patterns of life and patterns of death in the poorhouse population helped resolve conflicting documentary and skeletal evidence, and contributed to a more complete understanding of life and death in this nineteenth-century poorhouse. It is not surprising that this should be the case. Understanding human biology requires insight into human cultural issues, and understanding human cultural issues requires an understanding of human biology.

Modern correlates can be found to the nineteenth-century biocultural model that we have explored in this chapter. Brezinka and Padmos (1994), Hudelson (1996), Gijsbers van Wijk *et al.* (1996) McKinlay (1996), and Zaidi (1996) are just a few researchers who have recognized that both sex and gender role influence twentieth-century patterns of morbidity and mortality. Differing economic opportunities, the complexity of social networks, varying health care systems, and differences in ideology clearly impact the manifestation of disease as well as its relationship to death. As modern researchers, our recognition of the complexity of human interactions can guide us towards understanding the past. Similarly, exploring the past provides insight into the complex interaction between sex, gender role, health and disease in modern populations.

Acknowledgments

Many thanks to Debra Brown, Christine Engel, Theresa Jolly, Alexia Sabor, Julie Smentek, Paula Tomczak and Patrick Waldron for their participation in the initial demographic assessment, to Michele Buzon for assistance with the figures, and to Dr. R. A. Stuart for his comments. This research, and the endeavor to incorporate students into faculty research, was supported by the National Science Foundation under Grant No. SBR–9350256. Any opinions, findings, conclusions or recommendations expressed in this material are those of the authors and do not necessarily reflect the views of the National Science Foundation.

References

Armelagos GJ and McArdle A (1975) Population, disease, and evolution. In AC Swedlund (ed.), *Population Studies in Archaeology and Biological Anthropology, A Symposium. Memoirs of the Society for American Archaeology* **30**:1–10.
Brezinka V and Padmos I (1994) Coronary heart disease risk factors in women. *European Heart Journal* **15**(11):1571–84.
Brown J (1941) *The History of Public Assistance in Chicago 1833 to 1893.* Chicago: University of Chicago Press.
Buikstra J (1981) Mortuary practices, paleodemography and paleopathology: A case study from the Koster Site (Illinois). In R Chapman, I Kinnes, and K Randsborg (eds.), *The Archaeology of Death.* Cambridge: Cambridge University Press, pp. 123–32.
Buikstra JE and Ubelaker DH (Eds.) (1994) *Standards for Data Collection from Human Skeletal Remains.* Proceedings of a seminar at The Field Museum of Natural History, Organized by Jonathan Haas. Arkansas Archeological Survey Research Series Number 44. Fayetteville: Arkansas Archaeological Survey.
Cohen MN and Armelagos GJ (Eds.) (1984) *Paleopathology at the Origins of Agriculture.* Orlando: Academic Press, Inc.
Fox R (1980) *The Red Lamp of Incest.* New York: Dutton.
Gijsbers van Wijk CMT, Vliet KPV, and Kolk AM (1996) Gender perspectives and quality of care: towards appropriate and adequate health care for women. *Social Science and Medicine* **43**(5):707–20.

Goodman AH, May RL, and Meindl RS (1993) Response of bone and enamel formation to nutritional supplementation and morbidity among malnourished Guatemalan children. *American Journal of Physical Anthropology* **92**:37–51.

Grauer AL (1992) A piece of Chicago's past: an analysis of the Dunning Cemetery (Abstract). *American Journal of Physical Anthropology*, Supplement 14:83.

Grauer AL and McNamara EM (1993) Exploring childhood morbidity and mortality in the Dunning Cemetery skeletal population (Abstract). *American Journal of Physical Anthropology*, Supplement 16:97–8.

Grauer AL and McNamara EM (1995) A piece of Chicago's past: patterns of subadult mortality from the Dunning Poorhouse Cemetery. In AL Grauer (ed.), *Bodies of Evidence, Reconstructing History Through Skeletal Analysis*. New York: Wiley-Liss, pp. 91–103.

Grauer AL, McNamara EM, and Houdek DV (1995) Health, disease and gender in the Dunning Poorhouse Cemetery population (Abstract). *American Journal of Physical Anthropology*, Supplement 18:101.

Harzig C (1997) *Peasant Wives – City Women: From the European Countryside to Urban America*. Ithaca: Cornell University Press.

Houdek DV (1995) *A Look at the Lives of Chicago's Poor: A Skeletal and Historical Examination of Cook County Poorhouse (1851–1869)*. MA thesis, University of Chicago.

Hudelson P (1996) Gender differentials in tuberculosis: the role of socioeconomic and cultural factors. *Tubercle and Lung Disease* **77**(5):391–400.

Katz MB (1983) *Popery and Policy in American History*. New York: Academic Press.

Katz MB (1986) *In the Shadow of the Poorhouse*. New York: Basic Books.

Keene D (1989) *Final Report-Archaeological Investigations: Dunning Cemetery Site*. Chicago: Loyola University of Chicago, Archaeological Research Laboratory.

Keene D (1991) *Dunning Site-Archaeological Investigation* Report Number 3. Chicago: Loyola University of Chicago, Archaeological Research Laboratory.

Martin E (1987) *The Woman in the Lady: A Cultural Analysis of Reproduction*. Boston: Beacon Press.

McKinlay JB (1996) Some contributions from the social system to gender inequalities in heart disease. *Journal of Health and Social Behavior* **37**:1–26.

Stansell C (1986) *City of Women: Sex and Class in New York, 1789–1860*. New York: Alfred A. Knopf, Inc.

Stuart-Macadam P (1985) Porotic hyperostosis: representative of a childhood condition. *American Journal of Physical Anthropology* **66**:391–8.

Walker PL (1995) Biases in preservation and sexism in sexing: some lessons from historical collections for the paleodemographers. In SR Saunders and A Herring (eds.), *Grave Reflections: Portraying the Past through Cemetery Studies*. Toronto: Canadian Scholar's Press Inc., pp. 31–47.

Weatherford D (1986) *Foreign and Female: Immigrant Women in America, 1840–1930*. New York: Schocken Books.

Weiss KM (1972) On the systematic bias in skeletal sexing. *American Journal of Physical Anthropology* **37**:239–50.

Zaidi SA (1996) Gender perspectives and quality of care in underdeveloped countries: disease, gender and contextuality. *Social Science and Medicine* **439**(5):721–30.

11

Gender, health, and activity in foragers and farmers in the American southeast: implications for social organization in the Georgia Bight

CLARK SPENCER LARSEN

It is often assumed by anthropologists and non-anthropologists alike that, while theoretically an attractive idea, gender is largely inaccessible in archaeological and paleontological settings. Some argue that gender attribution is simply too ambiguous to contribute to our reconstructions and interpretations of past human behavior. Social scientists who are used to studying the living, find it difficult to imagine how gender can be reconstructed when the flesh and blood, and behavior of an individual cannot be observed. In archaeological contexts, Wylie (1991:31) notes that 'the very identification of women subjects and women's activities is inherently problematic; they must be reconstructed from highly enigmatic data.'

Contrary to this assertion, gender is, in fact, a highly visible part of the past. Since the publication of Conkey and Spector's (1984) pathbreaking article on the archaeology of gender, numerous gender-related works have appeared in the archaeological literature, facilitating a shift in how we think about people and their behavior in the past (see Wright 1996). Nowhere is the potential for illuminating key aspects of the human social condition more apparent than in studies of human skeletal remains where sex identification of individuals provides a window onto gender. The close connection between what can be observed in archaeological skeletons and inferences about gender has been referred to only infrequently in the archaeological literature. However, a growing bioarchaeological literature on subjects such as the areas of tooth use (Larsen 1985), dental health (Larsen 1983; Walker and Hewlett 1990; Lukacs 1996; Reeves 1997), infection (Grauer 1991; Hollimon 1991), and activity (Walker and Hollimon 1989; Molleson 1994) in past populations, indicates the importance of this line of investigation to anthropology.

An individual's sex derived from a skeletal series does not automatically correlate with gender, since gender is a social not a biological construct. However, the leap from sex determination to social identity and behavioral inference is not a long one. This is true because bones and teeth of past populations provide a retrospective picture of life history, not only of the

165

individual but of the population from which the individual is drawn (Larsen 1997). These life histories are encoded during the lifetime of the individual, potentially having different impacts on the sexes, owing to gender-specific cultural, social, and behavioral factors.

In this chapter, inferences are drawn about subsistence economy, the organization of food production, division of labor, and the structure and social organization of work, based on the study of a large series of prehistoric and contact era human skeletal remains from the southeastern United States Atlantic coast (present-day states of Georgia and Florida). These inferences are drawn from observations of several osseous indicators of health and activity. Comparisons are made between groups of female and male skeletons in order to illustrate gender processes in these populations. Two major adaptive shifts that took place in this region help to frame the discussion: (1) the shift from a lifeway based exclusively on foraging, hunting, gathering, and fishing, to one incorporating maize agriculture as a significant part of the economy; and (2) the increased production and consumption of maize by native populations that followed the arrival of Europeans and the establishment of Roman Catholic missions. While I have previously reported on these populations and adaptive shifts (Larsen and Harn 1994; Larsen and Ruff 1994), here, I address explicitly the interface between gender and biology using various skeletal parameters of health and behavior. In particular, I consider the differential impact of these adaptive shifts on females and males. Did the shift from foraging to farming result in relatively poorer health in females than males? When Europeans arrived on the scene in the sixteenth century, was the burden of deteriorating health and increasing workload felt differently by women and men in native societies? These are fundamental questions that need to be considered in this and other settings where diet and behavior change appreciably. The wider consideration of gender and human biology, and especially health and activity based on the study of human remains, has heretofore not been attempted for this region of the American Southeast. This chapter provides the opportunity to document the significant contribution of paleopathology and bioarchaeology to gender studies.

The setting

The focus of study is the mid-region of the Georgia Bight, a large embayment extending from Cape Hatteras, North Carolina, to Cape Canaveral, Florida (Figure 11.1). The region is dominated by a series of offshore barrier islands. Between these islands and the mainland coast is an abundance of marsh islands, tidal creeks, inlets, rivers, and sounds. Analysis of plants and animals

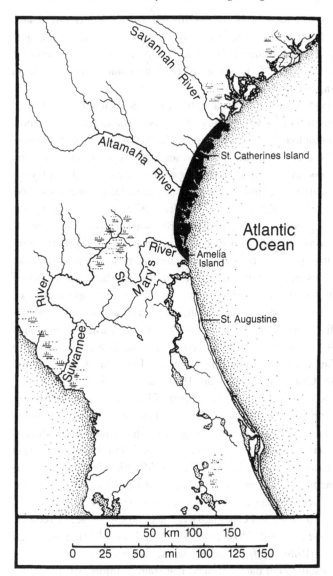

Fig. 11.1 Map of Georgia Bight, USA showing study region (shaded).

from archaeological sites representing the entire period of occupation of the region indicates that fishing and shellfish collecting were primary subsistence activities (Reitz 1988; Larsen *et al.* 1992).

There are three broad periods of native occupation in this region of the Georgia Bight. The oldest precedes the twelfth century AD, when populations subsisted primarily on foods collected or hunted from both terrestrial and marine settings. Populations were generally small, widely dispersed, and highly transitory. Human remains dating to this period are from mortuary sites dating from *c.* 1100 BC to the early twelfth century AD, with most remains postdating AD 700.

The second period is marked by the adoption of maize agriculture, occurring during the early to middle twelfth century. Analysis of stable isotopes (carbon and nitrogen) from human bone samples indicates that maize became a fundamental part of the diet. Although this represented a major reorientation of diet, the wild plants and animals previously exploited continued to play an important role in native nutrition. In some areas, particularly those associated with the primary ceremonial center at the Irene site, maize consumption actually declines appreciably in this period, perhaps as a response to drought, social stress, or some combination thereof (Larsen *et al.* 1992). Settlement data indicate that population nucleation and sedentism took place in some localities during this time, beginning in the twelfth and thirteenth centuries AD (e.g., Crook 1984). This period is also marked by the appearance of chiefdom levels of social organization, along with elaborate mound construction and symbolism characteristic of the late prehistoric Eastern Woodlands (Steponaitis 1986). Human remains utilized for this study are from a group of prehistoric sites dating between AD 1150 and 1550. Most are from the Irene Mound site (Caldwell and McCann 1941; Larsen 1982).

The third period commences with the arrival of European explorers during the middle sixteenth century and, of greater importance for native populations, the establishment of long-term missions beginning in the late sixteenth century. These missions were established among the Guale in the central Georgia Bight and other native groups in Spanish Florida (Milanich 1995). The northernmost and principal outpost on the Georgia coast was located on St. Catherines Island at Mission Santa Catalina de Guale (Thomas 1987; Larsen 1990). Archaeobotanical, isotopic, and elemental evidence indicates a heavy reliance on maize (Larsen *et al.* 1992; Ezzo *et al.* 1995) and use of some European plant domesticates (Ruhl 1990) by Indians at the mission.

Archaeological settlement data and historical sources inform us that populations became more nucleated and settled during the seventeenth century than during previous times, as groups moved to the major village associated with Mission Santa Catalina on St. Catherines Island. Historical accounts also indicate that the period was one of increasing stress due to European-introduced disease, poor nutrition, conflict, and generally deteriorating conditions.

The historical record provides an important perspective on physical activity and behavior during the seventeenth century. In this region – as is the case with other provinces of New Spain – the *repartimiento* draft labor was in full vogue (Hann 1988; Weber 1992). Able-bodied adults, mostly men, were required to contribute labor for a variety of activities, such as building projects, transport of materials over lengthy distances, road construction, and even military service. This labor often resulted in the displacement of men from their home villages. The compensation for these efforts was minimal, and the physical toll was high (Hann 1988; Larsen and Ruff 1994).

Following an attack on St. Catherines Island in 1680 by British and British-ally Indians, the native population fled southward, ending up on Amelia Island, Florida, by 1686. Called Santa Catalina de Guale de Santa Maria, the mission there enjoyed fewer than 20 years of occupation by native peoples. Its forced abandonment occurred in 1702 as the British continued their southward expansion. Archaeological, ethnohistoric, isotopic, and other analyses indicate a picture of human adaptation and settlement similar to that of St. Catherines Island.

For the contact period, human remains are from Santa Catalina on St. Catherines Island and its descendant population on Amelia Island. These remains represent an early and a late period of occupation in the contact era.

The population samples are divided into four groups of skeletons, precontact preagricultural, precontact agricultural, early contact, and late contact. A summary of the sites yielding human remains is presented in Table 11.1.

Methods of paleopathological study

Three pathological conditions are examined as indicators of health, stress, and activity: dental caries; periostitis; and osteoarthritis. Dental caries is a disease process characterized by focal demineralization of dental hard tissues by organic acids produced by bacterial fermentation of dietary carbohydrates, especially sugars (Newbrun 1982; Larsen 1997). In archaeological teeth, the lesions resulting from cariogenesis range in size from minute holes to extensive cavitation and loss of tooth structure. The disease causing the lesions is multifactorial, but presence of plaque, a composite of complex indigenous oral bacteria (e.g., *Streptococcus mutans*, *Lactobacillus acidophilus*), salivary glycoproteins, and inorganic salts adhering to tooth surfaces, is essential. Consumption of maize, a dietary carbohydrate containing sugar, is an important consideration for the setting of the Georgia Bight (Larsen *et al.* 1991). In this regard, consumption of carbohydrates can be expected to result in increased caries activity and appearance of carious lesions. Indeed, paleopathologists have identified an increase in frequency of dental caries in the

Table 11.1. *Mortuary localities, Georgia Bight, USA*

Site code	Site	N (1413)	Reference
Precontact preagricultural (pre-AD 1150)			
101	South New Ground Mound	1	Thomas and Larsen (1979)
102	Cunningham Mound C	4	Thomas and Larsen (1979)
103	Cunningham Mound D	2	Thomas and Larsen (1979)
104	Cunningham Mound E	1	Thomas and Larsen (1979)
105	McLeod Mound	14	Thomas and Larsen (1979)
106	Seaside Mound I	17	Thomas and Larsen (1979)
107	Seaside Mound II	8	Thomas and Larsen (1979)
108	Evelyn Plantation	3	Larsen (1982)
109	Airport	54	Larsen (1982)
110	Deptford (nonmound)	47	Larsen (1982; DePratter (1991)
111	Walthour	2	Larsen (1982; DePratter (1991)
112	Cannons Point	18	Larsen (1982)
113	Cedar Grove, Mound B	2	Larsen (1982); DePratter (1991)
114	Cedar Grove, Mound A	1	Larsen (1982); DePratter (1991)
115	Sea Island Mound	33	Larsen (1982)
116	Johns Mound	65	Larsen and Thomas (1982)
117	Marys Mound	5	Larsen and Thomas (1982)
118	Charlie King Mound	15	Larsen (1982)
119	Cedar Grove Mound C	8	Larsen (1982); DePratter (1991)
120	South End Mound II	25	Larsen and Thomas (1986)
121	Indian King's Tomb	5	Waring (1977); Larsen (1982)
Precontact agricultural (AD 1150–1500)			
201	North End Mound	1	Moore (1897)
202	Low Mound, Shell Bluff	1	Moore (1897)
203	Townsend Mound	2	Moore (1897); Cook (1970)
204	Deptford Mound	5	Larsen (1982); DePratter (1991)
205	Norman Mound	25	Larson, LH (1957); Larsen (1982)
206	Kent Mound	25	Cook (1978); Larsen (1982)
207	Lewis Creek, Mound II	7	Cook (1966); Larsen (1982)
208	Lewis Creek, Mound III	10	Cook (1966); Larsen (1982)
209	Lewis Creek, misc.	3	Cook (1966); Larsen (1982)
	Lewis Creek, Mound E	2	Cook (1966); Neighbors and Rathbun (1973, unpublished data); Sexton and Rathbun (1977, unpublished data)
210	Red Knoll	5	Larsen (1982)
211	Seven Mile Bend	18	Larsen (1982); Larsen *et al.* (nd)
212	Oatland Mound	2	Cook and Pearson (1973); Larsen (1982)
213	Seaside Mound 2	2	Thomas and Larsen (1979)
214	Irene Mound	267	Hulse (1941); Larsen (1982)
215	Grove's Creek	2	Larsen *et al.* (1998)
216	South End Mound I	19	Moore (1897); Larsen *et al.* (nd)
217	Skidaway Mitigation 3	3	Larsen *et al.* (1998)
218	Little Pine Island	17	Larsen *et al.* (1998)
219	Red Bird Creek Mound	3	Pearson (1984); Larsen *et al.* (nd)

Table 11.1. (*cont.*)

Site code	Site	N (1413)	Reference
220	Couper Field	44	Wallace (1975); Zahler (1976); Larsen *et al.* (nd)
221	Taylor Mound	30	Wallace (1975); Zahler (1976); Cook and Pearson (1973); Larsen *et al.* (nd)
222	Indian Field	22	Wallace (1975); Zahler (1976);
223	Taylor Mound/Martinez Test B	2	Martinez (1975); Larsen *et al.* (1998)
Early contact (AD 1550–1680)			
301	Santa Catalina de Guale	335	Larsen (1990); Larsen (1993)
302	Pine Harbor	109	Cook (1988); Hutchinson and Larsen (1988); Larsen *et al.* (nd)
Late contact (AD 1686–1702)			
303	Santa Catalina de Santa Maria	122	Larsen (1993); Larsen *et al.* (nd)

Eastern Woodlands when maize was adopted by late prehistoric societies (Milner 1984; Larsen *et al.* 1991).

Periostitis is characterized by osseous plaques with demarcated margins or irregular elevations of bone surfaces. The skeletal tissue in the unhealed form of periostitis is loosely organized woven bone. In the healed form, osseous tissue is incorporated into the normal cortical bone and the surface is frequently smooth, undulating, and somewhat inflated. In archaeological series, these lesions are generally isolated occurrences present on a single part of one skeletal element. Periostitis is a basic inflammatory response to bacterial infection, but localized trauma – such as a blow to the shins – may also be a causative factor as both conditions will stimulate the subperiosteal layer to produce bone (Eyre-Brook 1984; Simpson 1985). Most periostitis in archaeological populations is probably due to infection. Because of the generally unknown specific cause of periostitis – which bacteria and under what circumstances – it is considered to be a nonspecific indicator of infection (Larsen 1997). Infection appears to be a common condition for human populations living in sedentary or crowded conditions. Analyses of archaeological skeletal samples from settings in eastern North America (e.g., Goodman *et al.* 1984) to medieval urban Britain (Grauer 1991, 1993) and other contexts (see Larsen 1997) show relatively high levels of periostitis.

Osteoarthritis is a multifactorial disorder affecting articular joints, representing a range of responses to various predisposing factors (Hough and Sokoloff 1989; Rogers and Waldron 1995). The skeletal changes associated with the disorder include proliferative exophytic growths of new bone on joint margins (called 'osteophytes' or 'lipping') and/or erosion of bone on joint surfaces. In severe degenerative arthritis where the cartilage separating the joint surfaces of two bones is missing entirely, the articular surfaces become polished because of direct bone-on-bone contact (called 'eburnation').

A variety of factors causing osteoarthritis has been identified, including hormones, nutrition, metabolism, infection, trauma, and heredity (McCarty and Koopman 1993). The evidence based on living populations, especially in relation to physical demands of specific occupations (see Larsen 1997, for review) and rural versus urban behaviors (e.g., Jordan *et al.* 1995), indicates that mechanical stress figures prominently in the etiology of osteoarthritis (Hough and Sokoloff 1989; Larsen 1997). Occasionally, the general pattern and degree of physical activity – ranging from minor to excessive – can be identified in archaeological human remains with some success (e.g., Merbs 1983). The relationship between mechanical stress and osteoarthritis, however, is complex (Waldron 1994).

In making diachronic and synchronic comparisons for this study, the presence or absence of the three pathological conditions was recorded. Carious lesions were identified on all deciduous and permanent teeth. Periostitis was identified in all major long bones of juvenile and adult skeletons, but given its greater involvement in the lower leg relative to other parts of the body, only the tibia is discussed here. Osteoarthritis was recorded for adults for the major articular joints when either type of degenerative change, proliferative or destructive, was present. (See Larsen *et al.* 1991, 1996, nd; Larsen and Harn 1994 for detailed discussion of methods of data collection for caries, periostitis, and osteoarthritis.) Biological sex and age were identified using a standard protocol outlined in Bass (1995). Individuals 16 years or older were considered adult. Individuals younger than 16 years were considered juvenile.

Population trends in pathology

Dental caries

Change in dental caries prevalence in the Georgia Bight is marked by a general increase through time, with the exception of a slight decline in the early contact group (Table 11.2) Tooth by tooth comparisons indicate that most of the temporal variation reflects an increase in caries in the posterior dentition, specifically the premolars and molars, teeth that have complex

Table 11.2. *Adult dental caries prevalence, Georgia Bight, USA*

	PP[a] % (n)[b]	PA % (n)	EC % (n)	LC % (n)	Significant change[c]
Females					
Maxilla					
I1	0.0 (48)	3.7 (82)	8.6 (23)	31.4 (35)	EC/LC
I2	0.0 (39)	6.0 (66)	8.0 (25)	31.5 (38)	EC/LC
C	0.0 (52)	17.0 (103)	5.4 (37)	29.7 (37)	PP/PA,EC/LC
P3	0.0 (61)	21.0 (100)	17.5 (40)	34.4 (29)	PP/PA
P4	0.0 (61)	14.4 (114)	5.2 (38)	47.5 (40)	PP/PA,EC/LC
M1	0.0 (73)	16.7 (144)	20.9 (43)	51.5 (33)	PP/PA,EC/LC
M2	0.0 (77)	18.3 (129)	7.1 (42)	65.5 (29)	PP/PA,EC/LC
M3	0.0 (73)	17.4 (109)	12.5 (32)	79.1 (24)	PP/PA,EC/LC
Mandible					
I1	0.0 (33)	0.0 (58)	4.0 (25)	7.6 (39)	–
I2	0.0 (42)	2.4 (84)	0.0 (27)	20.4 (44)	EC/LC
C	0.0 (60)	5.1 (98)	7.3 (41)	29.5 (44)	EC/LC
P3	0.0 (65)	8.1 (124)	11.3 (44)	35.8 (39)	PP/PA,EC/LC
P4	0.0 (76)	13.9 (135)	10.8 (37)	51.5 (33)	PP/PA,EC/LC
M1	1.3 (79)	26.8 (131)	29.0 (31)	61.9 (21)	PP/PA,EC/LC
M2	1.2 (86)	31.5 (131)	20.0 (35)	84.2 (19)	PP/PA,PA/EC,EC/LC
M3	1.1 (92)	26.1 (117)	24.2 (33)	80.9 (21)	PP/PA,EC/LC
Total	1.2 (1017)	15.2 (1725)	12.3 (553)	41.9 (525)	PP/PA,EC/LC
Males					
Maxilla					
I1	2.1 (37)	0.0 (63)	4.7 (21)	12.8 (39)	PA/EC
I2	0.0 (31)	1.7 (58)	0.0 (25)	12.1 (41)	–
C	0.0 (35)	4.9 (82)	6.4 (31)	20.4 (49)	–
P3	0.0 (39)	27.6 (88)	12.5 (32)	26.0 (50)	PP/PA
P4	0.0 (38)	13.4 (89)	13.3 (30)	30.4 (46)	PP/PA
M1	0.0 (46)	18.5 (94)	21.1 (33)	42.4 (33)	PP/PA
M2	0.0 (47)	13.5 (89)	29.4 (34)	63.6 (33)	EC/LC
M3	0.0 (40)	16.1 (81)	30.5 (36)	57.1 (35)	PP/PA,EC/LC
Mandible					
I1	0.0 (15)	0.0 (58)	5.2 (19)	7.6 (39)	–
I2	0.0 (24)	1.4 (70)	4.7 (21)	25.5 (44)	EC/LC
C	0.0 (42)	0.0 (85)	0.0 (26)	22.4 (49)	EC/LC
P3	0.0 (43)	3.2 (93)	6.0 (33)	29.4 (51)	EC/LC
P4	0.0 (41)	8.0 (92)	3.0 (33)	45.4 (44)	EC/LC
M1	2.1 (47)	22.0 (100)	30.0 (23)	78.5 (28)	PP/PA,EC/LC
M2	2.1 (47)	12.9 (85)	37.0 (27)	90.3 (31)	PP/PA,EC/LC
M3	2.1 (45)	22.9 (98)	39.2 (28)	65.7 (35)	PP/PA,EC/LC
Total	0.6 (617)	10.9 (1325)	15.9 (452)	36.3 (647)	PP/PA,PA/EC,EC/LC

(Adapted from Larsen *et al.* 1991)

[a]PP: precontact preagricultural; PA: precontact agricultural; EC: early contact; LC: late contact.
[b]Number of teeth observed for presence/absence of carious teeth.
[c]Comparisons showing statistically significant change; chi-square: $p \leq 0.05$, two-tailed.

surfaces susceptible to cariogenesis. This increase is consistent with the general observation that once maize is adopted by late prehistoric eastern North American societies, caries prevalence increases (see Milner 1984; Larsen *et al.* 1991). With regard to the Georgia Bight, there are moderate increases following the introduction of maize, but an especially striking increase occurs later, in the late contact population from Amelia Island.

Both sexes show an increase in frequency of dental caries. However, the increase in females is more pronounced than in males (Figure 11.2). Female/male tooth type comparisons indicate that males are clearly less affected by dental caries than females, except in the molars during the early contact period. Combining all tooth types reveals that females in the precontact agricultural group have significantly more carious teeth than males (15.2% vs. 10.9%). The same pattern is noted in females in the late contact group (41.9% vs. 36.3%) (chi-square: $p \leq 0.05$).

Comparisons of archaeological populations worldwide associated with different times and settings reveal a common pattern of greater caries prevalence in females than males (see review in Larsen 1997). In most of these settings, and in the Georgia Bight in particular, the primary difference in behavior identified between sexes is in the amount of carbohydrates consumed, suggesting that the women and men had different dietary patterns, with women consuming more plant carbohydrates than men. This suggestion is supported

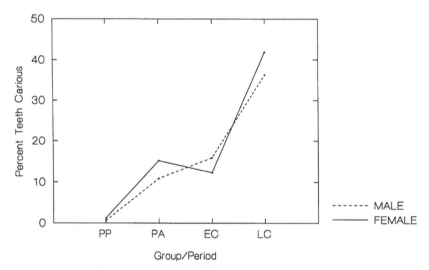

Fig. 11.2 Sex differences in caries prevalence, all tooth types combined, Georgia Bight. PP: precontact preagricultural, PA: precontact agricultural, EC: early contact, LC: late contact.

by differences observed in female and male subsistence behavior among historic and present-day traditional foragers and farmers. In the Eastern Woodlands of North America, Indian women were observed to have been responsible for most plant gathering and agricultural activities such as planting, harvesting, and especially for food preparation (see Swanton 1942; Hudson 1976; Watson and Kennedy 1991). Men were responsible for hunting, which was their primary contribution to the acquisition of dietary resources. This division of labor and food provision suggests that greater access to and consumption of maize by women resulted in their having higher caries rates.

Ethnographic documentation of dietary practices in late twentieth-century groups native to South America provides additional insight into sex differences in dental health (Walker *et al.* 1998). In three groups studied by Walker and coworkers (1998), the Yanomamo of Venezuela, the Yora of southeastern Peru, and the Achuar of Ecuador, meat and fish provide a significant part of the diet, but most carbohydrates are drawn from manioc, a cariogenic carbohydrate. Yora and Achuar women have greater exposure to manioc, owing largely to their practice of processing it in their mouths for beer (*chicha*). Not surprisingly, these women have a greater prevalence of dental caries than men.

Given the generally higher frequency of caries in females than males in these various settings, could there be some non-behavioral factor involved? Permanent teeth erupt at slightly earlier ages in females than males, thus exposing their teeth to cariogenic factors for a longer period of time. However, tooth eruption differences between females and males show inconsequential correlations with dental caries prevalence in modern populations (e.g., Moorrees 1957). There is a conventional belief that pregnancy compromises dental health, provoking caries and tooth loss. However, such a relationship is not borne out by scientific evidence (see Walker and Hewlett 1990; Larsen *et al.* 1991).

If sex-related caries differences are not the result of behavioral factors, then greater prevalence of caries in women than men should be ubiquitous, or nearly so. In fact, my search of the literature reveals a number of important exceptions showing either equal or greater prevalence of caries in males than females (see Larsen 1997).

Findings from the Georgia Bight suggest that differing subsistence activities and the division of labor between women and men resulted in different patterns of oral health. Although dental consequences of increased agriculture were experienced by all, the women in Georgia Bight societies, prehistoric and late contact, bore the greater brunt of the dental health costs associated with intensive carbohydrate consumption.

Table 11.3. *Tibial Periostitis Prevalence, Georgia Bight, USA*

	PP[a] %	(n)[b]	PA %	(n)	EC %	(n)	LC %	(n)	Significant change[c]
Total[d]	9.5	(126)	19.8	(331)	15.4	(26)	59.3	(96)	PP/PA, EC/LC
Female	4.3	(47)	24.1	(133)	14.3	(7)	65.7	(35)	PP/PA, EC/LC
Male	9.3	(32)	23.6	(93)	23.1	(13)	70.0	(36)	PP/PA, EC/LC

(Adapted from Larsen et al. 1998)

[a]PP: precontact preagricultural; PA: precontact agricultural; EC: early contact; LC: late contact.
[b]Number of tibiae observed for presence/absence of periostitis.
[c]Comparisons showing statistically significant change; chi-square: $p \leq 0.05$, two-tailed.
[d]Total: all juveniles and adults in the groups.

Periostitis

Comparisons of periostitis prevalence on the tibia in the four periods are presented in Table 11.3. These comparisons reveal that for the total population, there is an increase in periostitis from the earliest to the latest period, with a slight decline in the early contact sample, followed by a remarkable increase in the late contact period. This increase in the late prehistoric and late contact groups is consistent with increases identified in other settings in eastern North America (e.g., Lallo and Rose 1979; Milner 1991; Larsen 1995). It is not clear why these increases occurred. Given the presence of an endemic, non-venereal form of syphilis in eastern North America prior to and after contact (see Powell 1990), it is possible that some of the infectious pathology is due to that disease. However, most workers agree that the increases in periostitis in eastern North America are likely due to the increase in sedentism, population nucleation, and density accompanying the adoption of agricultural lifeways (Lallo and Rose 1979; Larsen 1995). In sedentary settlements, with reduced abilities to dispose of trash, conditions are established that are conducive to the spread and maintenance of infectious disease. Disease transmissibility is also facilitated by close contact between individuals. This would appear to be the case for late prehistoric Georgia Bight populations. In the contact samples, especially in the late contact group from Amelia Island, the populations were encouraged, and in some instances forced, to relocate to the mission. By concentrating populations, mission priests were able more efficiently to control native groups (Hann 1988). The health costs, insofar as they are reflected by the presence of periostitis, appear to have been high for these groups.

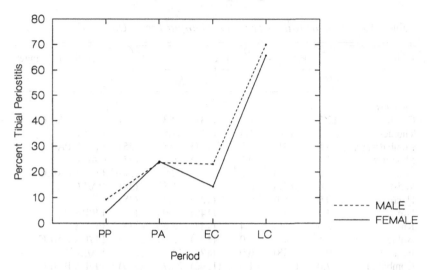

Fig. 11.3 Sex differences in tibial periostitis prevalence, Georgia Bight, USA. PP: precontact preagricultural, PA: precontact agricultural, EC: early contact, LC: late contact.

Unlike the case with dental caries, the difference between periostitis in adult females and males is not obvious (Figure 11.3). None of the sex differences identified for the four groups are statistically significant (chi-square: $p > 0.05$). However, in the contact period samples there is a greater prevalence in males than in females. Although the sex differences are not statistically significant, they may be important, especially in view of the differences in some behaviors of women and men living in missions. It is possible that exposure to new pathogens resulting from contact with other populations may be an important consideration. Historic records reveal the use of native laborers in regions far from their home villages, for example in St. Augustine, the capital of Spanish Florida (e.g., Hann 1988). The accounts note the return of laborers to their home villages following the completion of assigned labor. These laborers, having been exposed to new pathogens, may have served as agents for the transport of infectious disease to their home villages and missions. These records also inform us of the predominance of men in the transported labor force (Hann 1988). Thus, the greater prevalence of males among individuals with skeletal infections may reflect their greater exposure to novel pathogens.

Table 11.4. *Osteoarthritis prevalence, Georgia Bight, USA*

Joint	PP[a]		PA		LC		Significant change[c]
	%	n[b]	%	n	%	n	
Females							
Cervical	17.2	(29)	1.4	(73)	42.3	(26)	PP/PA,PA/LC
Thoracic	6.7	(30)	1.4	(72)	57.5	(33)	PA/LC
Lumbar/sacral	19.5	(51)	9.9	(111)	54.5	(55)	PP/PA,PA/LC
Shoulder	2.4	(83)	0.7	(144)	56.0	(25)	PA/LC
Elbow	9.6	(94)	0.0	(167)	2.9	(34)	PP/PA,PA/LC
Wrist	2.6	(77)	0.0	(140)	5.3	(38)	PA/LC
Hand	0.0	(50)	0.8	(129)	5.3	(38)	–
Hip	4.3	(93)	0.0	(148)	0.0	(36)	PP/PA
Knee	15.0	(94)	3.4	(147)	8.6	(35)	PP/PA
Ankle	4.5	(88)	0.0	(139)	15.6	(32)	PP/PA,PA/LC
Foot	0.0	(48)	0.0	(120)	11.1	(27)	PA/LC
Combined	7.0	(737)	1.5	(1390)	23.7	(379)	PP/PA,PA/LC
Males							
Cervical	40.0	(20)	11.3	(53)	44.4	(27)	PA/LC
Thoracic	12.5	(16)	11.8	(51)	65.4	(26)	PA/LC
Lumbar/sacral	34.6	(26)	16.3	(80)	52.9	(51)	PP/PA,PA/LC
Shoulder	10.5	(38)	1.7	(120)	11.1	(27)	PP/PA,PA/LC
Elbow	13.7	(51)	6.1	(114)	10.3	(29)	PP/PA
Wrist	2.6	(39)	0.9	(106)	10.0	(30)	PA/LC
Hand	0.0	(28)	2.0	(100)	3.2	(31)	–
Hip	0.0	(51)	9.1	(110)	6.4	(31)	–
Knee	18.6	(59)	12.6	(111)	7.4	(27)	–
Ankle	4.1	(49)	9.2	(109)	10.0	(30)	–
Foot	0.0	(26)	1.1	(93)	41.6	(24)	PA/LC
Combined	9.0	(403)	6.9	(1047)	30.5	(302)	PP/PA,PA/EC

(Adapted from Larsen *et al.* 1996)

[a]PP: precontact preagricultural; PA: precontact agricultural; EC: early contact; LC: late contact.
[b]Number of articular joints observed for presence/absence of osteoarthritis.
[c]Comparisons showing statistically significant change; chi-square: $p \leq 0.05$, two-tailed.

Osteoarthritis

Group comparisons in osteoarthritis prevalence are presented in Table 11.4. These comparisons reveal a clear pattern of reduction in osteoarthritis accompanying the shift from foraging to farming in the pre-contact periods. Significant reductions are present for the lumbar/sacral, shoulder, and elbow joints for males and the cervical, lumbar/sacral, elbow, hip, knee, and ankle joints

Table 11.5. *Individuals affected by osteoarthritis per five year age group, Georgia Bight, USA*

Age	PP[a]		PA		LC		Significant change
	%	n[b]	%	n	%	n	
16.1–20	0.0	(9)	6.7	(30)	33.3	(3)	–
20.1–25	11.1	(18)	5.7	(35)	0.0	(3)	–
25.1–30	10.0	(10)	9.1	(11)	60.0	(5)	PA/LC[c]
30.1–35	0.0	(1)	42.9	(7)	25.0	(8)	–
35.1–40	66.7	(6)	20.0	(10)	78.6	(14)	PP/PA[c],PA/LC[d]
40.1–45	25.0	(4)	33.3	(3)	81.0	(21)	–
45.1+	46.7	(15)	62.5	(8)	85.7	(21)	–
Total	23.8	(63)	15.4	(104)	69.3	(75)	PA/LC[c]

(Adapted from Larsen *et al.* 1996)

[a]PP: precontact preagricultural; PA: precontact agricultural; EC: early contact; LC: late contact.
[b]Number of articular joints observed for presence/absence of osteoarthritis.
[c]$p \leq 0.1$, Fisher's Exact Test (two-tailed).
[d]$p \leq 0.01$, chi-square (with Yates's Correction for Continuity).
[e]$p \leq 0.0001$, chi-square.

for females (chi-square: $p \leq 0.05$). However, in the late contact group (comparisons are not available for the early contact group), prevalence increases dramatically, especially in the male cervical, thoracic, lumbar/sacral, shoulder, wrist, and foot joints, and the female cervical, thoracic, lumbar/sacral, shoulder, elbow, wrist, ankle, and foot joints, all of which show statistically significant increases (chi-square: $p < 0.05$).

Osteoarthritis is strongly influenced by the age structure of a population – the older the individual, the greater the chance of having the disorder (Hough and Sokoloff 1989). Indeed, the precontact agriculturalists have a lower mean age at death than the precontact preagriculturalists (19.7 years vs. 26.1 years), and the late contact group has the highest mean age at death (29.8 years). Thus, based on age structure of these three groups, the observed changes in prevalence of osteoarthritis could easily be predicted. In order to test this hypothesis, individuals with osteoarthritis were distributed into five-year age categories (Table 11.5). These comparisons reveal that proportionately fewer individuals were affected in the precontact agriculturalists than in the precontact preagriculturalists, and likewise, the proportion of individuals affected by osteoarthritis greatly increases in the late contact group. Therefore, although the changing frequencies of articular joints showing osteoarthritic remodeling may be related in part to the changing age structure of the skeletal samples, the prevalence of osteoarthritis between periods reflects a real rather than a superficial change.

It can be concluded that the reversal in osteoarthritis prevalence in the late contact group indicates that the mission population from Amelia Island engaged in activities that placed heavy loading on the articular joints of the skeleton. The interpretation of reduction in workload and activity in the shift from foraging to farming and increase in workload and activity in the contact period is supported by biomechanical analysis of femora and humeri (Ruff *et al.* 1984; Ruff and Larsen 1990; Larsen and Ruff 1994; Larsen *et al.* 1996). That is, analysis of cross-sectional geometric properties that reflect bone strength show a decrease in strength from the precontact preagricultural period to the precontact agricultural period and an increase from the early contact to the late contact periods. Although bone strength may not be caused by the same factors or behaviors that cause osteoarthritis, the overall picture of changing activity as shown by osteoarthritis and mechanical analysis are similar.

With regard to sex differences in osteoarthritis, males display more osteo-arthritis than females (Figure 11.4), almost certainly reflecting the greater physical demands upon males than upon females (Larsen *et al.* 1995). Interest-ingly, females show a greater number of statistically significant increases in the late contact sample relative to the precontact agriculturalists (Table 11.4).

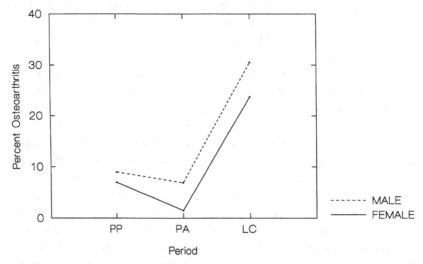

Fig. 11.4 Sex differences in osteoarthritis prevalence, all articular joints combined, Georgia Bight, USA. PP: precontact preagricultural, PA: precontact agricultural, LC: late contact.

This may suggest that the missions, and the associated behavioral changes, had a greater impact on women than on men. However, comparisons of osteoarthritis prevalence in females and males in the late contact sample show that the frequencies for many of the joints are more similar than in earlier periods. For example, the intervertebral joints have very similar prevalence values for both sexes. This decreased sexual dimorphism of osteoarthritis suggests that differences in gender-based work and other activity became negligible in the contact period. This finding is supported by the decreased sexual dimorphism found in the bilateral asymmetry of humeral cross-sections in the contact periods relative to the prehistoric periods (Fresia *et al.* 1990; Larsen 1997). Like the prevalence for osteoarthritis, these structural bone changes indicate that the activities of women and men – at least as they affect the articular joints and bone structure – became more similar in the mission settings of the Georgia Bight. We are not able to identify particular activities engaged in by males and females in Spanish missions. However, one historical account remarks that maize preparation, an arduous physical activity involving the use of the arms and upper body, was a responsibility of males (Hann 1986). Because osteoarthritis increases in both females and males, it is possible that the activities of both women and men were labor-intensive, involving similar kinds of physical effort. Perhaps, too, as agriculture became more intensive during the contact period, women and men shared the responsibilities of maize production by doing the same or similar activities.

Summary and conclusions

The findings of this study can be summarized as follows. Between the precontact preagricultural and the late contact periods: (1) there is an overall increase in dental caries prevalence, with a greater increase in women than in men; (2) there is an overall increase in periostitis prevalence, with somewhat greater prevalence in late contact males than in late contact females; and (3) there is a reduction in osteoarthritis in prehistoric farmers relative to foragers, followed by a dramatic increase in the contact or mission period, with an appreciable decrease in sexual dimorphism in the late contact period. The greater caries prevalence in women than men suggests that women had a greater intake of carbohydrates than men, perhaps as a result of their greater involvement in food preparation. The increases in periostitis prevalence suggest a deterioration in living conditions, especially among the contact period populations. Nearly all of the lesions are well healed, suggesting that individuals affected represent those who may have more successfully mitigated pathogenic agents that cause bony responses than those who lack bone infections altogether. This interpretation runs counter to what we know about living conditions of the missions, however, especially with regard to increasing population aggregation and declining health.

The somewhat higher prevalence of periostitis in adult males in the late contact period may reflect a more robust response to infection on the part of males. Alternatively, this chapter argues that the greater frequency in males may be due to the travel of adult males to other areas of Spanish Florida in the *repartimiento* system, and consequently, exposure to new pathogens not present at home villages.

Despite gender variation in conditions resulting in poorer health, the similarities in osteoarthritis between sexes indicate increased similarity in physical activities that cause the disorder, such as heavy labor associated with agricultural production. The decreased differences in osteoarthritis prevalence in women and men in the late contact period indicate that the types and levels of work and activity may have become more similar and more strenuous.

Pathology and biomechanical evidence have been used in this study to examine the variation in response of women and men to changing circumstances, these include major shifts in subsistence and settlement and the arrival of Europeans and establishment of mission centers. I believe that the findings of this study have significance for the study of both past and living human populations. This study shows that change in subsistence and activity, including different types of behaviors associated with food production and consumption, can have very real consequences for women and men alike. However, the consequences can be differential. We have seen, for example, that women in the Georgia Bight possibly had greater access to, and consumed more, domestic plant carbohydrates, resulting in relatively poorer dental health than men. Similarly, in living populations, greater access to plant carbohydrates in women in traditional societies leads to poorer dental health (e.g., Walker and Hewlett 1990). Likewise, the study of living populations and the different roles of women and men provides essential information for improving our understanding of pathology variation in past groups. The fact that traditional Yora and Achuar women in South America chew manioc for beer preparation and that they have more carious lesions than males helps us to understand the higher frequencies of carious lesions in southeastern United States native populations.

This study also points to the importance of using a number of different indicators of health and lifeway when studying past and living populations. Focus on single indicators of behavior and dietary change presents an overly narrow perspective on human adaptation. A comprehensive understanding of gender must be grounded in a comprehensive analytical approach.

Acknowledgments

I thank Anne Grauer and Patty Stuart-Macadam for the invitation to participate in the symposium organized for the 1995 meetings of the American Association of Physical Anthropologists held in Oakland, California, and for the opportunity to contribute to this volume. Marianne Reeves commented on an earlier draft. This chapter is a contribution to the *La Florida Bioarchaeology Project*. Funding for the research came from the National Science Foundation (BNS–8406773, BNS–8703849, SBR–9305391).

References

Bass W (1995) *Human Osteology: A Laboratory and Field Manual*, 4th edn. Columbia: Missouri Archaeological Society.

Caldwell J and McCann C (1941) *Irene Mound Site, Chatham County, Georgia*. Athens: University of Georgia Press.

Conkey M and Spector J (1984) Making the connections: feminist theory and archaeologies of gender. In MB Schiffer (ed.), *Advances in Archaeological Method and Theory*. New York: Academic Press, pp. 1–38.

Cook FC (1966) *The 1966 Excavations at the Lewis Creek Site*. Research Manuscript No. 274. Athens: University of Georgia, Laboratory of Archaeology.

Cook FC (1970) *The 1970 Excavation at the Townsend Mound*. Research Manuscript No. 275. Athens: University of Georgia, Laboratory of Archaeology.

Cook FC (1978) *The Kent Mound: A Study of the Irene Phase on the Lower Georgia Coast*. MA thesis, Florida State University, Tallahassee.

Cook FC (1988) Archaeological evidence for the distribution of sixteenth-century Irene/Guale people on the Georgia coast and relationships to socio-political organization. *Early Georgia* **16**:1–27.

Cook FC and Pearson CE (1973) *Three Late Savannah Burial Mounds in Glynn County, Georgia*. Research Manuscript No. 276. Athens: University of Georgia, Laboratory of Archaeology.

Crook MR, Jr. (1984) Evolving community organization on the Georgia coast. *Journal of Field Archaeology* **11**:247–63.

DePratter CB (1991) *W.P.A. Archaeological Excavations in Chatham County, Georgia: 1937–1942*. Laboratory of Archaeology Series, Report. No. 29. Athens: University of Georgia.

Eyre-Brook AL (1984) The periosteum: its function reassessed. *Clinical Orthopaedics and Related Research* **189**:300–7.

Ezzo JA, Larsen CS, and Burton JH (1995) Elemental signatures of human diets from the Georgia Bight. *American Journal of Physical Anthropology* **98**:471–81.

Fresia AE, Ruff CB, and Larsen CS (1990) Temporal decline in bilateral asymmetry of the upper limb on the Georgia coast. In CS Larsen (ed.), *The Archaeology of Mission Santa Catalina de Guale: 2. Biocultural Interpretations of a Population in Transition. Anthropological Papers of the American Museum of Natural History* **68**:121–32.

Goodman AH, Lallo J, Armelagos GJ, and Rose JC (1984) Health changes at Dickson Mounds, Illinois (AD 950–1300). In MN Cohen and GJ Armelagos (eds.), *Paleopathology at the Origins of Agriculture.* Orlando: Academic Press, pp. 271–305.

Grauer AL (1991) Life patterns of women from medieval York. In D Walde and ND Willows (eds.), *The Archaeology of Gender.* Archaeological Association of the University of Calgary, Proceedings of the 22nd Annual Chacmool Conference, pp. 407–13.

Grauer AL (1993) Patterns of anemia and infection from medieval York, England. *American Journal of Physical Anthropology* **91**:203–13.

Hann JH (1986) The use and processing of plants by Indians of Spanish Florida. *Southeastern Archaeology* **5**:91–102.

Hann JH (1988) *Apalachee: The Land Between the Rivers.* Gainesville, University of Florida Press.

Hollimon SE (1991) Health consequences of divisions of labor among the Chumash Indians of southern California. In D Walde and ND Willows (eds.), *The Archaeology of Gender.* Archaeological Association of the University of Calgary, Proceedings of the 22nd Annual Chacmool Conference, pp. 462–9.

Hough AJ and Sokoloff L (1989) Pathology of osteoarthritis. In DJ McCarty (ed.), *Arthritis and Allied Conditions*, 11th edn. Philadelphia, Pennsylvania: Lea & Febiger, pp. 1571–94.

Hudson C (1976) *The Southeastern Indians.* Knoxville: University of Tennessee Press.

Hulse FS (1941) The people who lived at Irene: physical anthropology. In J Caldwell and C McCann (eds.), *Irene Mound Site, Chatham County, Georgia.* Athens: University of Georgia Press, pp. 57–68.

Hutchinson DL and Larsen CS (1988) Determination of stress episode duration from linear enamel hypoplasias: a case study from St. Catherines Island, Georgia. *Human Biology* **60**:93–110.

Jordan JM, Linder GF, Fryer JG, and Renner JB (1995) The impact of arthritis in rural populations. *Arthritis Care and Research* **8**:242–50.

Lallo JW and Rose JC (1979) Patterns of stress, disease and mortality in two prehistoric populations from North America. *Journal of Human Evolution* **8**:323–35.

Larsen CS (1982) *The Anthropology of St. Catherines Island:* 3. *Prehistoric Human Biological Adaptation. Anthropological Papers of the American Museum of Natural History* **57**, Part 3.

Larsen CS (1983) Behavioural implications of temporal change in cariogenesis. *Journal of Archaeological Science* **10**:1–8.

Larsen CS (1985) Dental modifications and tooth use in the western Great Basin. *American Journal of Physical Anthropology* **67**:393–402.

Larsen CS (Ed.) (1990) *The Archaeology of Mission Santa Catalina de Guale:* 2. *Biocultural Interpretations of a Population in Transition. Anthropological Papers of the American Museum of Natural History* **68**.

Larsen CS (1993) On the frontier of contact: mission bioarchaeology in La Florida. In BG McEwan (ed.), *The Spanish Missions of La Florida.* Gainesville: University Press of Florida, pp. 322–56.

Larsen CS (1995) Biological changes in human populations with agriculture. *Annual Review of Anthropology* **24**:185–213.

Larsen CS (1997) *Bioarchaeology: Interpreting Behavior from the Human Skeleton.* Cambridge: Cambridge University Press.

Larsen CS and Harn DE (1994) Health in transition: disease and nutrition in the Georgia Bight. In KD Sobolik (ed.), *Paleonutrition: The Diet and Health of Prehistoric Americans. Southern Illinois University at Carbondale, Center for Archaeological Investigations, Occasional Paper* **22**:222–34

Larsen CS and Ruff CB (1994) The stresses of conquest in Spanish Florida: structural adaptation and change before and after contact. In CS Larsen and GR Milner (eds.), *In the Wake of Contact: Biological Responses to Conquest.* New York: Wiley-Liss, pp. 21–34.

Larsen CS and Thomas DH (1982) *The Anthropology of St. Catherines Island*: 4. *The St. Catherines Period Mortuary Complex. Anthropological Papers of the American Museum of Natural History* **57**, Part 4.

Larsen CS and Thomas DH (1986) *The Anthropology of St. Catherines Island*: 5. *The South End Mound Complex. Anthropological Papers of the American Museum of Natural History* **63**, Part 1.

Larsen CS, Crosby AW, Griffin MC, Hutchinson DL, Ruff CB, Russell KF, Schoeninger MJ, Sering LE, Simpson SW, Takacs JL, and Teaford MF (nd) A biohistory of health and behavior in the Georgia Bight:1. The agricultural transition and the impact of European contact. In RH Steckel and JC Rose (eds.): *The Backbone of History: Health and Nutrition in the Western Hemisphere.* (In press.)

Larsen CS, Ruff CB, and Griffin MC (1996) Implications of changing biomechanical and nutritional environments for activity and lifeway in the eastern Spanish borderlands. In BJ Baker and L Kealhofer (eds.), *Bioarchaeology of Native American Adaptation in the Spanish Borderlands.* Gainesville, Florida: University Press of Florida, pp. 96–125.

Larsen CS, Ruff CB, and Kelly RL (1995) Structural analysis of the Stillwater postcranial human remains: behavioral implications of articular joint pathology and long bone diaphyseal morphology. In CS Larsen and RL Kelly (eds.), *Bioarchaeology of the Stillwater Marsh: Prehistoric Human Adaptation in the Western Great Basin. Anthropological Papers of the American Museum of Natural History* **77**:107–33.

Larsen CS, Schoeninger MJ, van der Merwe NJ, Moore JM, and Lee-Thorp JA (1992) Carbon and nitrogen stable isotopic signatures of human dietary change in the Georgia Bight. *American Journal of Physical Anthropology* **89**:197–214.

Larsen CS, Shavit R, and Griffin MC (1991) Dental caries evidence for dietary change: an archaeological context. In MA Kelley and CS Larsen (eds.), *Advances in Dental Anthropology.* New York: Wiley-Liss, pp. 179–202.

Larson LH (1957) The Norman Mound, McIntosh County, Georgia. *Florida Anthropologist* **10**:37–52.

Lukacs JR (1996) Sex differences in dental caries rates with the origins of agriculture in South Asia. *Current Anthropology* **37**:147–53.

Martinez CA (1975) *Culture Sequence on the Central Georgia Coast, 1000 BC–1650 AD* MA thesis, University of Florida, Gainesville.

McCarty DJ and Koopman WJ (Eds.) (1993) *Arthritis and Allied Conditions*, 12th edn. Philadelphia: Lea & Febiger.

Merbs CF (1983) *Patterns of Activity-Induced Pathology in a Canadian Inuit Population.* National Museum of Man Series. Archaeological Survey of Canada, Mercury Series Paper No. 119. Ottawa: National Museums of Canada.

Milanich JT (1995) *Florida Indians and the Invasion from Europe.* Gainesville: University Press of Florida.

Milner GR (1984) Dental caries in the permanent dentition of a Mississippian period population from the American Midwest. *Collegium Anthropologicum* **8**:77–91.

Milner GR (1991) Health and cultural change in the late prehistoric American Bottom, Illinois. In ML Powell, PS Bridges, and AMW Mires (eds.), *What Mean These Bones?* Studies in Southeastern Bioarchaeology. Tuscaloosa: University of Alabama Press, pp. 52–69.

Molleson T (1994) The eloquent bones of Abu Hureyra. *Scientific American* **271**(2):70–5.

Moore CB (1897) Certain aboriginal mounds of the Georgia Coast. *Journal of the Academy of Natural Sciences of Philadelphia* **11**.

Moorrees CFA (1957) *The Aleut Dentition: A Correlative Study of Dental Characteristics in an Eskimoid People.* Cambridge, Mass.: Harvard University Press.

Newbrun E (1982) Sugar and dental caries: a review of human studies. *Science* **217**:418–23.

Pearson CE (1984) Red Bird Creek: late prehistoric material culture and subsistence in coastal Georgia. *Early Georgia* **12**:1–39.

Powell ML (1990) On the Eve of Conquest: Life and Death at Irene Mound, Georgia. In CS Larsen (ed.), *The Archaeology of Santa Catalina De Guale: 2. Biocultural Interpretations of a Population in Transition. Anthropological Papers of the American Museum of Natural History* **68**:26–35.

Reeves M (1997) Sexual dimorphism in caries rates and the Creek division of labor. *American Journal of Physical Anthropology*, Supplement 24:194–5.

Reitz EJ (1988) Evidence for coastal adaptation in Georgia and South Carolina. *Archaeology of Eastern North America* **16**:137–58.

Rogers J and Waldron T (1995) *A Field Guide to Joint Disease in Archaeology.* Chichester: John Wiley & Sons.

Ruff CB and Larsen CS (1990) Postcranial biomechanical adaptations to subsistence changes on the Georgia coast. In CS Larsen (ed.), *The Archaeology of Mission Santa Catalina de Guale: 2. Biocultural Interpretations of a Population in Transition. Anthropological Papers of the American Museum of Natural History* **68**:94–120.

Ruff CB, Larsen CS, and Hayes WC (1984) Structural changes in the femur with the transition to agriculture on the Georgia coast. *American Journal of Physical Anthropology* **64**:125–36.

Ruhl DL (1990) Spanish mission paleoethnobotany and culture change: a survey of the archaeobotanical data and some speculations on aboriginal and Spanish agrarian interactions in La Florida. In DH Thomas (ed.), *Columbian Consequences*, vol. 2. *Archaeological and Historical Perspectives on the Spanish Borderlands East.* Washington, DC: Smithsonian Institution Press, pp. 555–80

Simpson AHRW (1985) The blood supply of the periosteum. *Journal of Anatomy* **140**:697–704.

Steponaitis VP (1986) Prehistoric archaeology in the southeastern United States, 1970–1985. *Annual Review of Anthropology* **15**:363–404.

Swanton JR (1942*) Source Material on the History and Ethnology of the Caddo Indians.* Bureau of American Ethnology, Bulletin 132.

Thomas DH (1987) *The Archaeology of Mission Santa Catalina de Guale: 1. Search and Discovery. Anthropological Papers of the American Museum of Natural History* **63***, Part 2.*

Thomas DH and Larsen CS (1979) The Anthropology of St. Catherines Island: 2.

The Refuge-Deptford Mortuary Complex. Anthropological Papers of the American Museum of Natural History **56**, Part 1.

Waldron T (1994) *Counting the Dead: The Epidemiology of Skeletal Populations.* Chichester: John Wiley & Sons.

Walker PL and Hewlett BS (1990) Dental health diet and social status among Central African foragers and farmers. *American Anthropologist* **92**:382–98.

Walker PL and Hollimon SE (1989) Changes in osteoarthritis associated with the development of a maritime economy among southern California Indians. *International Journal of Anthropology* **4**:171–83.

Walker PL, Sugiyama L, and Chacon R (1998) Diet, dental health, and cultural change among recently contacted South American Indian hinter-horticulturalists. In JR Lukacs (ed.), *Dental Anthropology.* Albert A. Dahlberg Memorial Volume. (In press.)

Wallace RL (1975) *An Archeological, Ethnohistoric, and Biochemical Investigation of the Guale Aborigines of the Georgia Coastal Strand.* PhD dissertation, University of Florida, Gainesville.

Waring AJ, Jr. (1977) The Indian King's tomb. In S Williams (ed.), *The Waring Papers: The Collected Works of Antonio J. Waring, Jr.* Papers of the Peabody Museum of Archaeology and Ethnology, Harvard University **58**:209–15.

Watson PJ and Kennedy MC (1991) The development of horticulture in the Eastern Woodlands of North America: women's role. In JM Gero and MW Conkey (eds.), *Engendering Archaeology: Women and Prehistory.* Oxford, England: Basil Blackwell, pp. 255–75.

Weber DJ (1992) *The Spanish Frontier in North America.* New Haven: Yale University Press.

Wright RP (1996) Introduction: gendered ways of knowing in archaeology. In RP Wright (ed.), *Gender and Archaeology.* Philadelphia: University of Pennsylvania Press, pp. 1–19.

Wylie A (1991) Gender theory and the archaeological record: why is there no archaeology of gender? In JM Gero and MW Conkey (eds.), *Engendering Archaeology: Women and Prehistory.* Oxford, England: Basil Blackwell, pp. 31–54.

Zahler JW, Jr. (1976) *A Morphological Analysis of a Protohistoric-Historic Skeletal Population from St. Simons Island, Georgia.* MA thesis, University of Florida, Gainesville.

Index

Page numbers for figure materials are in bold.